D1798824

Martin Nicholas Kunz

best | designed

wellness hotels

WESTERN AND CENTRAL EUROPE . ALPS . MEDITERRANEAN

avedition lebensart

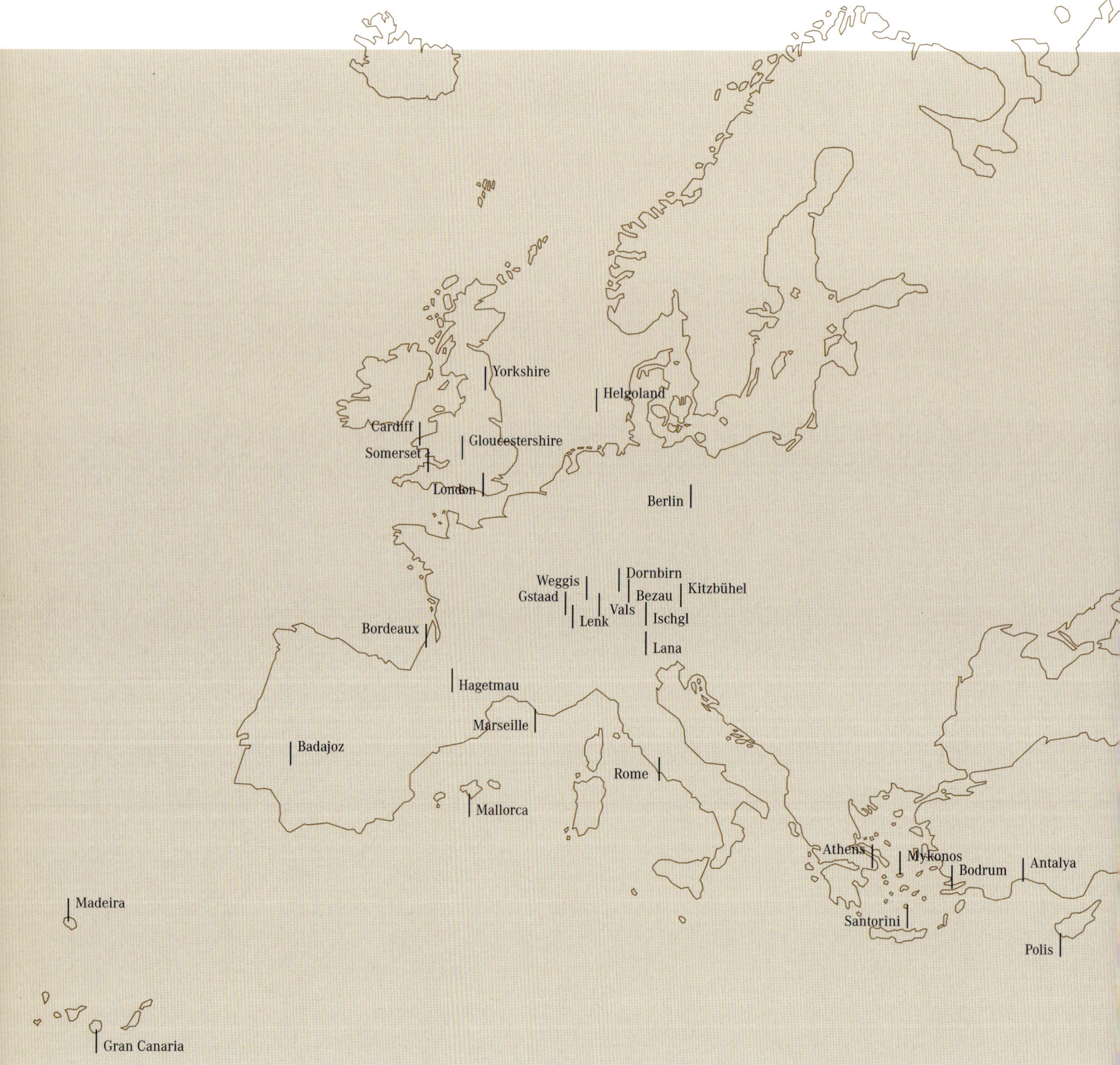

01 02

Another morning in another anonymous hotel. After all, who wants to wake to sights, sounds and scents that evoke a sense of place? Scents such as the enticing aroma of fresh croissants from the bakery next to the hotel. Or sounds such as the twitchy hooting of excitable taxi drivers. If only you could open the windows. Wouldn't it be nice to have a little balcony, or at least windows that reach the floor, so that you could greet the morning sun, allow a breath of Paris, London, New York into the room or simply drink in the silent landscape spread out before you. But what's the point of individuality in a hotel, why should a place have its own unique character? As long as the price is acceptable, the bed comfortable and you can find your way around blindfold no matter where in the world you are. Who cares, as long as there is a hairdryer next to the mirror, a television tucked away behind the doors of the fake wood cabinet and a telephone next to the toilet.

Planning hotels must be one of the most rewarding design commissions, and yet creative mediocrity continues to prevail. With an unerring eye for poor taste, designers fiddle about with the decor, producing a cacophony of pseudo gold and ostentatious marble and happily using colour schemes that agonisingly just fail to match. And worse still, "but it's practical" is the excuse routinely used to cleverly conceal every trace of functionality. Starting with the location of the power sockets. Which frequent traveller has never crawled under a writing desk in the hunt for a socket and unplugged the mini-bar by mistake? And we won't even mention the poetry of architecture.

The view of "what guests really want" still seems to be astonishingly uniform within the international hotel industry: champagne buckets and flower arrangements have become tediously predictable

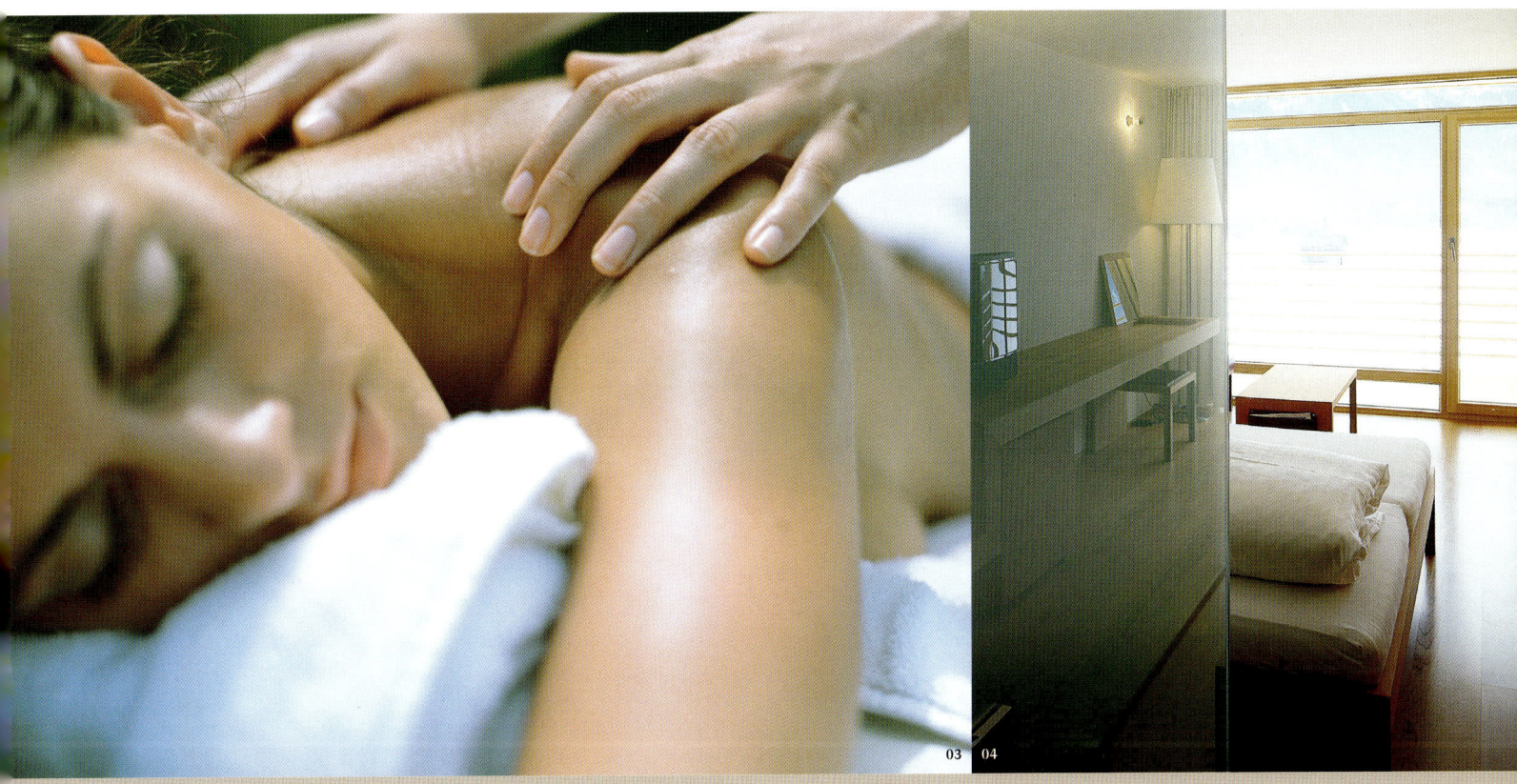

03 04

shorthand for hospitality and comfort wherever you go. Closely followed by "practical" uniformity: management consultants would call it CI. "So that guests immediately feel at home no matter where in the world they are". And never know where they are when they wake up in the morning.

Well no, not quite. Like that little village in Gaul led by Asterix and Obelix, there is a small but growing group of hotels which, ever since the 1980s, has been fighting back against this trend with a mix of traditional, gastronomic values and radical new interpretations, to cater for a new type of traveller. Tellingly, perhaps, most of the

proponents of this development do not hail from within the hotel industry. Many are people who travel often and enjoy doing so. Who, every time they stay in a hotel, note what is missing and work out how they could do it better. At some point, everything they have observed and registered comes together and it is then that they begin turning the dream into reality.

Frequent business travellers in the creative professions (fashion, design, architecture, communication, and entertainment), as well as discerning individual travellers are consciously seeking out smaller, owner-managed hotels to

01 | Radical minimalism in the Thermal Vals by Peter Zumthor.

02 | Wellness Hotel concept of the new generation in the Vigilius Mountain Resort by Matteo Thun.

03 | More and more hotels have substantial offers of wellness but still too rarely in the surroundings of good architecture.

04 | Post Hotel in Bezau by Oskar Leo and Johannes Kaufmann.

05 06 07

escape the monotonous sameness of the chains. There are now perhaps a few hundred such hotels around the world where architects and interior designers have deliberately focused on the concept of "home from home" in their interpretations. The results of their efforts are very different – from the starkest minimalism to out-and-out kitsch. From private holiday home which the owner would rather share with paying guests than be bored alone, to the reawakening of once grand hotels. The common denominators are the aim of the designers and owners to create something artistic and individual, and the desire of the guests to feel that they are spending time in a special place with special people.

These hotels are consciously different from the overwhelming majority. Above all, they are worlds

apart from the identikit chains. The only thing they have in common is that they have virtually nothing in common, at least as far as looks, architecture and interior design are concerned. But there is more to it than just visual modernity. This type of hotel is also a rejoinder to rituals that are apparently set in stone. For example, why should breakfast end at eleven o'clock or – worse still – ten o'clock in the morning? Hotel professionals can reel off 100 reasons why this has to be so. But there are hundreds of thousands of guests who would like to be able to have their breakfast at two in the afternoon, even mid-week. And more and more of the new breed of hoteliers are showing that it's perfectly possible. More often than not the unthinkable does actually work. Seeking out the unusual, the bold and sometimes the surprising

was one of the reasons for collecting the hotels for this book.

Highly individual hotel concepts based on actual experience of travelling, in combination with brilliant architecture, help to restore inner balance. And that's even before you start adding sauna, massage, steamroom or swimming pool. If these ingredients are present too, the product is a holistic wellness experience catering for all the senses. Clearly, aesthetics is a very subjective issue. The traveller who is at his happiest as he glides into the water between the smoothly polished granite walls of the Vals thermal baths will either have an inborn penchant for radical puritanism, or will have been deeply interested in architecture and design for many years. This type of architecture is simply beyond the range of

08   09

experience of many uninitiated visitors. The total absence of any form of decoration can leave you reeling. Just as a potpourri of materials and colours can easily send your brain into overdrive. As in the Palm Beach Hotel on Gran Canaria, for example, or in the operetta-style lobby in the Sanderson. While over-the-top decoration can be enjoyed every now and then as a bit of fun, radical minimalism can sometimes be simply liberating. Assuming, of course, that you allow yourself the pleasure of getting to grips with it.

The other examples collected here all have their own unique story to tell; each one has its individual merits and special features. Whether you are gazing at the ocean from the pool at the Estalagem da Ponta do Sol or at Potsdamer Platz from the Grand Hyatt indoor pool, lounging behind canvas around on a sofa in the Belvedere on Mykonos or counting the minarets of Bodrum on the terrace of the Marmara, one thing is for certain. This is not a dream.

Martin Nicholas Kunz

05 | Relaxation on the cliffs. Estalagem da Ponta do Sol in Madeira.

06 | Swimming pool without an edge in Maricel in Mallorca.

07 | Meditation behind cloister walls in the Hospederia Convento de la Parra.

08 | Sanus per aquam – healthy with water is the wellness credo of the 21st century.

09 | Building by Ryan de Matos Storey at the country estate of Cowley Manor in Gloucestershire.

choupana hills resort & spa | madeira . portugal

DESIGN: Michel de Camaret, Didier Lefort

Those accustomed to flat countryside and straight roads would be advised to either let themselves be chauffered around or to start off gently by climbing the less daunting hills. Even powerful cars can only manage certain ascents in first gear, and the descents require very reliable brakes. Funchal's streets, or alleyways to be exact, easily put those of San Francisco in the shade and promise a thrilling, white-knuckle experience. To reach Madeira's newest luxury resort, such test rides can be enjoyed to the full. Up high, still a way above the already imposing, famous botanical gardens, on a completely exposed plateau and in the midst of lush, sub-tropical vegetation, investors have built a new style resort and wellness hotel, with 34 elegantly designed houses perched on stilts.

Early in the planning stage, the group of architects, Frenchmen Michel de Camaret and Didier Lefort, drew up an ecological concept that would entail integrating the surrounding nature as a fundamental component. The wood used for fixtures and furniture originates from Asia. Its use is balanced out by an extensive forestry programme for the tropical rain forest.

At the heart of the resort is an extensive range of health, beauty and wellness offers, with massage techniques from around the globe, including hydromassages. Aromatherapy is also available, as are revitalising baths, sauna, Turkish hammam, face and body treatments, as well as diets or set nutrition plans. The architecture and design take on a significant meaning, in the knowledge that the aesthetic surroundings promote a sense of well-being and enhance stimulation.

The buildings, internal areas, furniture, decoration and plants are interwoven with the external areas such as pathways, garden, swimming pool or terraces. The harmonious architecture reveals a depth of feeling, and mixes all kinds of elements from the history of Portuguese art with Asian forms and contemporary internationalism. Tropical woods and coconut mats, for example, are juxtaposed with fine linen cloth and smooth, uniformly painted walls. Buddha statues stand alongside futuristically elegant bars and steel/wood constructions; Tibetan antiques share space with contemporary seating interpretations and bar stools. Almost all of the furniture is designed by Didier Lefort. His signature already marks the furnishings of the Datai on the island of Langkawi (Malaysia).

Except for the four suites, all the villas on wooden stilts – referred to as bungalows – are divided into two 40m² units with living and sleeping quarters, an airy bathroom flooded with natural

02 03   04

05

02 | Hammam.

03 | Across 34 mini villas: 60 rooms
     and 4 suites with a living space of
     up to 40 m² and a sun veranda.

04 | The interior architecture is purist but
     dominated by warm woods.

05 | View of Lounge terrace of the main
     building at the pool.

daylight, as well as an extensive terrace made of wooden boards. Two full-length sliding glass doors can be fully opened and form a seamless link between the internal area and the terrace, which covers an area of almost 30m². From here the panoramic view over Funchal towards the ocean lying far below is sensational. This view can also be enjoyed from the main buildings, drawing on fine Asian architecture, with their reception, lobby, various terraces, wellness centre with indoor pool, bar and restaurant. Those swimming a few lengths in the green shimmering lagoon-pool will undoubtedly keep pausing, namely whenever they take in the view of the blue ocean, merging visually with the flat edge of the overflowing water.

06 | Indoor Lounge.

07 | Reception and Lobby.

08 | The influence of Asian architecture
     cannot be overseen.

crowne plaza resort | madeira . portugal

DESIGN: Ricardo Nogueira, Duarte Caldeira Silva

A pair of cheerily nattering white bathrobes walk down the long, cool corridor, and enter the lift at its end. The press of a button, a short wait, the doors open – the bathrobes move forward into the cabin. Almost like peering out from a lighthouse, the rear glass wall affords a spectacular view of the ocean. Out on the shorefront, the couple make their way to the two huge, deep-blue saltwater pools, and prepare themselves for the first, exciting exploratory plunge.

The remarkable volume of these geometrically laid out pools, bare of any adornment, is indicative of the size of the resort, and its organisation as a wellness and conference hotel. With 300 rooms

and suites, it is the newest major project on Madeira in the five-star segment and is not typically representative of the other, mostly smaller, buildings presented in this book. The concept, however, is certainly interesting. The private investor and owner, an established lover of art and architecture, commissioned two of his fellow countrymen, Ricardo Nogueira and Duarte Caldeira Silva, to plan both the complex and his private house, situated immediately beside it.

Even when driving by, along the "Estrada Monumental" coastal road – the name of the road is perfectly apt – the colossus towers up in the foreground. In contrast to other large

hotels, the modernity and transparency of the construction is immediately striking, created, to an extent, by extensive glazing. The two identically arranged halves of the building form an open, bright effect, a seemingly appropriate response to the concrete blocks and almost criminally bad architecture of bygone days. Optically, the façade is particularly attractive at dusk when the interior is lit. The blue colour of the corridor walls on all 10 floors radiates through the glass fronts, bathing the façade in an eerily atmospheric glow.

While the rooms are practical and promise a comfortable stay, complete with Philippe Starck

01 | 02

01 | Atrium of the left building wing. The halls
in the ten-storey hotel offer a view onto the
town.

02 | Glass lends the rather monumental building
a certain lightness.

03 04

05

03 | D oor detail in one of the suites on the 10th floor.

04 | The sport and wellness area contains two saltwater pools...

05 | ... and an indoor pool.

06 | In total there are 276 rooms and 24 suites.

07 | Chaise-longue with sea view in a suite. The standard rooms are rather interchangeable from a design point of view.

06 07

furnished balconies and sea views, they appear rather less innovative compared with the hotel's architecture and the design of the internal public areas. They are reminiscent of American luxury chains which, although not to be sneered at in the hotel business as a whole, seldom produce original results.

The rooms would certainly have benefited from the use of lighter materials, more minimalism and more daylight – especially in the bathrooms. That said, the property's "Crowne Plaza Resort" label is somewhat misleading in a positive sense, projecting an image not quite congruent with its uncommon architectonic concept and service standards. The interior design of the breakfast hall and lobby remains compelling, situated in the right and left wings of the building, beneath the entrance area, in a museum-like atrium spread over three floors. Intensified by the brightness of the glass fronts, these areas appear very airy and give the impression of sitting in an enormous outdoor covered art gallery – especially as the doors to the terraces, situated in front, are mostly wide open. Immediately next to this are the two main restaurants; the more exclusive "Wild Orchid" with its fusion cuisine and "La Brasserie" with its light French cuisine. Those preferring fresh fish can eat in the simpler "Cervejara Portuguesa" on the pool deck.

The main attraction is most certainly the huge spectrum of wellness treatments on offer at the "Marine Spa", including a complete range of thalasso therapy. More standard facilities include the sauna, steam baths and indoor pool. Features such as the two squash courts, the kids club and the hotel's own diving platform, are certainly worth a special mention. With its striking architecture and impressive guest portfolio, the resort is a contemporary, modernist answer to the traditional grand hotel.

estalagem da ponta do sol | madeira . portugal

DESIGN: Tiago Oliveira

If the water was just a couple of degrees warmer, swimmers would probably emerge with wrinkled fingers and toes, as if from a long, leisurely bath. The swimming pool counts among those places at Estalagem da Ponta do Sol where one could stay forever, gazing out in fascination. Luckily, the temperature is cool enough that, after a few laps, returning to a warm sun lounger doesn't seem like an altogether bad idea.

A good 100 metres above steep cliffs, the view swings along a rocky coast, over the ocean, terraced hills and green, subtropical mountains, before coming to chic, reserved architecture. The complex, by the young Portugese architect, Tiago Oliveira, unifies all the aesthetic, usefulness, care, poetry and intelligence that one often wishes from architecture, but seldom experiences. This recent member of the "design hotels" group is a place for the discovery of beauty, and is well-suited to a study of the harmonious bond between nature and architecture.

The drive from Funchal takes around 30 minutes; from the airport it's around 45. The motorway leads through tunnels and over bridges, and finally along the stony shoreline. In the distance, the new construction crowns a projecting cliff with a beam of white. One last tunnel, a roundabout, and the vista suddenly changes. A bridge is the optical centrepoint, high up, linking the new building to the lift tower of the main structure – an old "Quinta", a Portuguese estate, with living quarters once belonging to the local lord.

A more spectacular location is difficult to imagine. All of the hotel's sections divide themselves along slate outcrops that tumble down to the sea. The two-storey main building, with reception, lobby, bar and clubroom, stands elevated above Ponta do Sol, encompassed by a roofed terrace and gardens with ancient trees and palms. Sinking into one of the wicker chairs, drinking a café com leite and enjoying the panorama of blue on blue – that's true luxury.

One floor, or better said, one layer of rock further up, Tiago Oliveira has placed an aesthetic highlight: the restaurant. The cuisine may be simple, almost rustic, but the setting is grandiose. A backdrop of surf, boulders, lush vegetation over rolling mountains and a fishing village that looks like it's been painted in oils. The edifice itself is a simple, white cube, glazed from floor to ceiling, with an introverted interior design. Its geometric clarity and renunciation of decorative elements makes guests' appreciation of the natural environment, as well as the architecture, somehow more intense. From the restaurant, a series of steps (or, less strenuously, a lift and

01 | The hotel complex sits on a cliff like a gem of modern architecture.

02 03

bridge) leads to the 54 guestrooms, characterised by their predominantly white walls and reduced furnishing.

Everything is kept to a minimum, but everything is there. The corridors run like bowed arcades, open to the air. Strolling down these elegant walkways, it is perfectly usual to see guests wander direct from their rooms to the pool, clad simply in slippers and a bath robe. All rooms have their own balcony or terrace, and those facing west have views across the village, hills and ocean. These rooms are also slightly larger than those that sit directly over the cliffs, towards the east. The easterly rooms are clearly more

popular, however, as they are also the most expensive; even though at Estalagem da Ponta do Sol, one cannot really talk about "expensive". With prices starting at around 100 Euro a night, the hotel counts among those secret tips that definitely offer extraordinary value for money.

There's no way to consider the Estalagem, other than as a concentrated charge of fascination, a real discovery – an almost unknown place of relaxation, revelation and inspiration. New pictures are formed every other moment, in the changing of the sunlight and the outlines of the clouds. The meditative, penetrating cymbal crash of waves on the battered coast becomes a mantra,

and all guests long to do is watch and listen, trying to catch the faint, fresh taste of sea salt on their tongue.

02 | From the outside one reaches the 54 rooms via steps and arcades.

03 | They are divide over two buildings with a view onto the fishing village of Ponta do Sol or directly over the cliffs.

04 | The most beautiful panoramic views is from the swimming pool.

05 | Restaurant.

08

09

06 | The "expelled" wellness area is not particularly large. This is not
necessary as the whole of the complex offers relaxation and
inspiration.

07 | One of the rooms over the cliffs. If you leave the balcony door open,
you can hear the whispering of the waves and crying of the gulls.

08 | Due to the lush vegetation Madeira is considered as the most desired
rambling paradise.

09 | Perfectly formed high seat above the Atlantic.

quinta da casa branca | madeira . portugal

DESIGN: João Favila V.S. Menezes, Teresa Gois Ferriera, Luís Rosário

Aside from the warm climate, another important aspect for the high standard of living in Funchal are the many green areas within the town itself. The town is a jumbled mix of close-ranked buildings, parks, copses, ravines, rock faces too steep to be built on, bare fields and huge private properties. Many of these were once managed estates, often banana plantations, in the hands of large Portugese or British landowners. The tourist boom was responsible for the bananas disappearing from the centre and outskirts of Funchal to make room for more hotels and guesthouses.

The Quinta da Casa Branca shares a similar history. The five-star property lies well hidden on a hill overlooking the ocean, in the middle of a vibrantly colourful park a mere 200 metres or so above the legendary 150-year-old Reid's Palace and roughly a 20-minute walk from the harbour

and town centre. The property is not quite so easy to find; its hidden driveway is reminiscent of the entrance to a grand villa. Behind the wrought-iron gate, however, is not the palatial building one would expect to see, but a dazzling filigreed glass house; a strictly geometric cube. The building won Joao Favila Menezes and his partner Maderia's most important architectural prize in 1999. One of the most impressive areas that shows the design to its best advantage, especially at dusk, is the terrace to the rear of the reception area, which affords a direct view of the ocean over the hills of the town, the hotel park and the noble private villas.

The 43 guest rooms, 12 superior and two suites, are located beneath the reception pavilion, with direct access to the garden, and in a further two-storey building, which was completed in 2002. This structure also houses a limited wellness

centre offering sauna, steam bath, fitness, jacuzzi and massage facilities. A charming peculiarity is the family Rolls Royce from the 1940s, which is used, complete with chauffeur, for ferrying guests around the local area.

The history of the property dates back to the mid-19th Century when the Scottish ancestors of the current owners, the Leacock family, cultivated wine and bananas on the land, then still covering an area of 6 hectares. The hotel, situated directly next to their house, still offers its guests, however, roughly 2 hectares of open green space with exotic plants, a well-heated pool, breakfast pavilion and a gourmet restaurant in one of the oldest buildings on the land.

It was the owner's openness and curiosity, together with his faith in the capability of the architects, who were friends of the family, that

01 | Reception and Lobby are located in a glass box à la Barcelona
Pavilion above the rooms.

02 | From the armchairs of the lobby one has a view over the hotel
park onto the ocean.

03 04

05

made the creation of this little jewel possible. It also led to a true architectonic rarity being created step-by-step from the initial mish-mash of styles in the interior design and garden furnishings; a process that is still ongoing. Joao Favila Menezes convinced his client with cool reason and clear sense: "My aim was to allow the powerful beauty of the natural environment to take effect using subdued, transparent archictecture. This was also the reason we primarily used glass and black stone."

The guests mostly still fit the Madeira cliché: distinguished, conservative and greying. With the combination of contemporary architecture and a young, fresh service concept, the still-young

French-Portuguese director, Isabel F. Ferraz, is however aiming towards an integrative change of generations: "Casa Branca is, after all, the hottest place on the island…," she says with a wink, adding "…as far as the weather's concerned, anyway."

03 | It is hard to imagine that the hotel is in the centre of Funchal.

04 | Façade elements of the bistro, a newly built area above the pool.

05 | Most of the rooms are located in a long narrow building.

06 | Couches in the Lobby.

07 | The guest rooms beneath the Lobby are accessible from the outside and inside.

08 | The building is situated in a two hectare sized park with an ancient collection of trees. From this point one has a panorama view over Funchal and the ocean.

09 | In an extension: new suites and the spa were built in 2002.

palm beach | gran canaria . spain

DESIGN: Alberto Pinto

In the 1970s, the thousand-year-old palm grove behind the Maspalomas dunes on Gran Canaria was developed into a quintessential sun, sea and palm tree holiday idyll. The Palm Beach Hotel, curvaceous and imposing, was built in that same decade. In autumn 2002, after several months of alterations, the hotel celebrated its reopening. All the hotel's spaces were completely redesigned by Paris-based architect and interior designer Alberto Pinto, using the same concept throughout. Pinto has liberated the original building from every superfluous detail, exposing the beauty of the place in a new way with touches that hark back to the style of the 1970s. He has breathed fresh life into this luxurious beach hotel while retaining its own special character.

The hotel's redesign was based on typical design elements of the period, such as loud colours,

dramatic contrasts and the use of materials like chrome, glass, lacquer, brass, mirror surfaces, travertine and marble. Pinto has taken these elements and developed them into a new stylistic concept characterised by clear lines. For the design of the 327 rooms, four colour palettes were chosen: the bold, sunny tones on the walls contrast with the shades of sand and sea reflected by the flooring. The unified room concepts in coral and blue, violet and green, yellow and turquoise convey a strong sense of colour and form. The rooms in beige and brown offer a subtle alternative. The Palm Beach Hotel's lobby and restaurants are modernistic, verging on retro, with colours in ultra-contrasting designs.

The hotel's wellness oasis is the thousand-year-old palm grove, laid out during the redesign work as a "health garden" with a health and wellness centre run by medical staff. A saltwater

pool with underwater massage was installed in addition to the heated freshwater pool that reflects sunlight in intricate turquoise mosaic patterns. Guests can tone up on fitness equipment shaded by the palm trees before enjoying a massage in the wellness centre. Anyone who tires of swimming in the Atlantic or in the two pools can choose between Oriental Rasul or Cleopatra baths, and saline or whirlpool baths with seaweed for a spot of sensual indulgence. The thalasso therapy and amazingly diverse types of saunas on offer – dry, steam, Turkish, Finnish and bio – are the ideal way to prepare the immune system for the inevitable return home to Central European climates.

01 | At home in the colours of the sea: Spatial concept in coral and blue.

02

03 04

05 06

02 | The quiet room in the wellness area with a view onto the
Japanese garden.

03 | The pool is situated in a thousand year old palm grove.

04 | Refreshing eyes and other senses: The sun and water mirror each
other in this room.

05 | Seating in the best position: only a step to the massage pool and
wellness centre.

06 | Light stone and Spanish tiles bring Mediterranean flair to the
bathroom.

hospederia convento de la parra | badajoz . spain

DESIGN: Francisco Wenao, Maria Ulecia

Central Spain, Badajoz to be more precise, south-west of Madrid on the border with Portugal. We are looking for a hotel called Hospederia Convento de la Parra (a mouthful even for Spaniards) when someone points us in the direction of the Convento, a former monastery on the outskirts of the city. Once we have found our treasure, we pause to take a deep breath and find that here, cradled in a landscape that can only be described as somewhere between rugged and gentle, time seems to have stood still.

Back in the 17th century, this was a place where monks lived the simple life; today it is a sanctuary where stressed 21st century visitors can find shelter and reconnect with their inner self. Behind the former cloister walls lies a complex of very different rooms waiting to be discovered. The hotel has 15 double rooms, four single rooms and two junior suites, as well as other areas that invite its guests to walk around, read, lie down, to dream. There is even a pool and a vegetable garden for those more comfortable outside. What is unique about this place is the complete absence of intrusive elements such as TV or radio – modern, technological achievements which others boast of as luxuries and to some seem indispensable in a life dominated by technical gadgetry. Rather than electronic distractions, you will find a library with countless books to delve into at your heart's content.

The interiors of all the rooms are decorated very subtly so as not to detract from the architectural effect of the old vaults, cloisters and halls. Strategically positioned lighting accentuates the structure and fabric of the building, giving a good impression of the effect the original builders intended to achieve. The rooms become more open the further you move away from the centre of the building: closed, intimate rooms at the heart of the monastery are surrounded by half-open covered walkways, which in turn give way to open courtyards. Here, your gaze is irresistibly drawn upwards into the deep blue Spanish sky. The secret to this authentic atmosphere is the combination of a few, natural materials such as terracotta, whitewashed plaster, iron and wood and the tried and tested language of form. Guests feel instantly secure and at ease – the best ingredients for a relaxing stay.

Those for whom the monastery walls are too claustrophobic are advised to go out into the surrounding countryside of Estremadura or the Peak Mountains, the Matamoras Valley or Monsalud. The historic towns of Feria, Zafra and Merida, with their well-preserved town centres, are also well worth a visit.

01 | Relaxation behind cloister walls. Absolute peace and inspiration are guaranteed.

02  03

04

02 | Instead of television, close vision. A wealth of details can be discovered.

03 | With a total of 19 rooms and 2 suites, the guests are amongst close company.

04 | Swimming pool in the interior court.

05

05 | The washing tables are reminiscent of past life in the cloister.

06 | Wonderful contrasts of white plush furniture and old walls.

07 | Relaxing in the arcade hall.

06 07

hotel maricel | mallorca . spain

DESIGN: Xavier Claramout, María Jesús Asiaín

Take the boat transfer from the airport to the Hotel Maricel, and you know you are in for an exclusive experience. The name "Maricel" is what's called a portmanteau word – a combination of the Spanish expressions for sea and sky. And sure enough, the imposing coastal villa does serve as a visual link between the two.

It was built in the 1940s by a wealthy merchant who wanted to accommodate his guests in high style. Soon this place became one of the most popular hang-outs for Mallorca's international scene and the site of extuberant parties and exclusive events. The structure is in effect a large, square tower with a broad wing facing the Mediterranean Sea. With vistas to rival any on the island, the hotel's terraces still have an air of jet-set luxury. A swimming pool close to the villa's pier extends right out into the sea.

Smooth transitions like these are what the Maricel is all about. Fresh design concepts are integrated into the original architecture, which still exudes the bourgeoisie charm of a bygone era. Traditional Balearic design elements of the original building have been brought back to life by architect Xavier Claramunt in the form of archways and arcades on several levels. The cool, dark interior is punctuated by the play of shadow and sunlight. Room dividers create cozy niches in the lounge. On the veranda just outside, rattan lounges have immaculate white upholstering.

The guestrooms were created with three basic principles in mind: comfort, space and light. Born of this philosophy Hotel Maricel's 29 double-rooms and suites are thoroughly modern. Furnishings have been selected for their clear lines and colours. Bathrooms are especially tasteful, outfitted with milkglass and simple fixtures. A freestanding bathtub is a throw back to earlier times. All in all, the feel at the Hotel Maricel is balanced somewhere between youthful elegance and a kind of stage-managed pomp.

01 | Banquettes and club interior made of rattan
on the veranda in Maricel.

02 | A table set in one of the numerous niches in
the building.

03

03 | Flowing transitions right into the architecture. The name "Maricel" stands for the melting of the sky into the sea.

04 | The Moorish pillars on the veranda frame the seductive look of a candlelight dinner.

04 05
06 07

05 | The Mediterranean cuisine of Maricel is certainly presented elegantly.

06 | A hint of past splendour. The baths are small, light paradises and play with the shimmering of body forms.

07 | Clear shapes, raw materials, selected interior. Fresh living feeling to relax in.

les sources de caudalie | bordeaux-martillac . france

DESIGN: Philippe Hurel

It seems the obvious thing to do: vino-therapy in one of the best wine regions in the world. In Graves, just 20 minutes by car south-east of Bordeaux, the world revolves around wine. Vines are an integral feature of the landscape of softly rolling hills that stretch as far as the eye can see. And in the middle of the vineyards stand the traditional, big-name wine-growing estates whose wines fetch top prices on the global market. Still, Florence and Daniel Cathiard did show a pioneering spirit when they put their money on the insights of a researcher at the University of Bordeaux who maintained that wine growers were throwing away the best bit, the grape pomace. Joseph Vercauteren, professor of pharmacology, discovered that grape seeds are full of polyphenols, natural plant metabolites that have antioxidant qualities. These substances fight free radicals, which are known to be the main cause of skin ageing.

The success story began in 1990 when Florence and Daniel, both former members of the French national skiing team and later extremely successful in marketing, were on the lookout for a new challenge and bought the run-down "Château Smith Haut Lafitte" estate. Under the suspicious eyes of their old-established neighbours, the newcomers were soon producing one of the region's best wines. But they weren't satisfied with that alone, so they built a hotel next to their estate.

The various houses grouped around the wine estate were built using old materials such as you find when old country houses are pulled down in southern France. The houses were given poetic names like "La Bastide des Grands Crus" and "La Grange au Bateau". They have wonderful wooden verandas and sun decks on which chirping birds and quacking ducks provide the

free musical accompaniment. And the "I'lle aux oiseaux' which sits on its stilts in the middle of a pond is a favourite with couples on honeymoon.

When thermal water enriched with valuable minerals was discovered 540 m under the earth, vino-therapy was born. All the therapies are based on the precious components in the grapes and their seeds which are said to detoxify and rejuvenate the body. It is a concept that has convinced famous faces like Isabelle Adjani and Donatella Versace. Peelings whose main ingredients are ground grape seeds and sea salt give skin a soft, velvety feel; bubbly baths in the barrique vat, traditionally used to store the region's wines, provide deep relaxation. And guests can drift away during the four-handed massage using fragrant natural essences.

01 | Swimming baths, treatment rooms and quiet z ones are located in this wooden house reminiscent of a large barn.

02

03

02 | The Spa is surrounded by a park in which the open-air swimming pool is situated. Speciality is the vino-therapy.

03 | La Tour des Cigares. Rustic leather ambience in smoker tower room.

04 | Part of the wellness arrangement is the well-considered decorative surroundings.

05 | The transition between the interior and exterior is marked by a glass wall which remains open on warm days.

06 | The planner placed a great deal of emphasis on the harmony of the working materals not only on such Scrabble exercisies.

07 | Quiet place in front of the swimming hall.

hôtel des lacs d'halco | hagetmau . france

DESIGN: Eric Raffy

In the heart of the Gascogne, not far from Chalosse, lies the Hotel des Lacs D'Halco. The region is a magnet for tourists who are drawn to the wide range of sports facilities for which it is renowned. Along with tennis, golf, riding and cycling, visitors can engage in all manner of water sports on one of the region's many waterways.

The aim of Eric Raffy, the architect in charge of planning, was to design a building which blends in well with the environment and at the same time tells a story. He took his inspiration from the surrounding terrain as well as the legend of the Lady of the Lake, creating a form of narrative architecture which embraces elements of the landscape and plays with them: The two-storey hotel building follows the bowed bank of the lake; the façade facing the lake is glass from floor to ceiling, so that guests inside the building get the

feeling they are still part of the landscape; the reflection of the building in the water gives the impression that it is melting with nature. A round offshore island floats on the water – surrounded and thus protected by the hotel complex and linked to it by a footbridge. The building's segmental arch focuses the eye on the floating rotunda that houses the restaurant and directs one's gaze towards the centre of the lake. Visitors thus either move in spaces that face the lake or that are on a tangent to it. The forms used to express this blending of edifice and nature had to be natural too, which is why right-angles are extremely rare in the ensemble of archways and sloping faces.

You first enter the building at lake-level. The parquet flooring has inlaid stone pathways leading to the most important areas in and around the hotel: the lake, the kitchen (which

is famous for its regional delicacies) and the pine forest. Next to and on the same level as the reception area lies the kitchen, the dining room overlooking the lake, a bar and a semicircular swimming pool. Stone stairways lead to the 24 rooms on the upper storey, half of which face the forest, half of which face out across the water.

The materials used (wood, stone, copper, glass) were chosen carefully because they pick up on features of the immediate environment, entering into a form of dialogue with the landscape. Wood was used in building the visible structure of the hotel, namely its supports and girders; the outer supports are reminiscent of the trunks of pine trees in the neighbouring forest. The grey-green of the copper-covered roof blends with the colours of the forest and the water. The glass fronts on both storeys slope inwards,

01 | The two-storey hotel building follows the curve of the riverbed. The glass front looks completely out onto the lake.

02 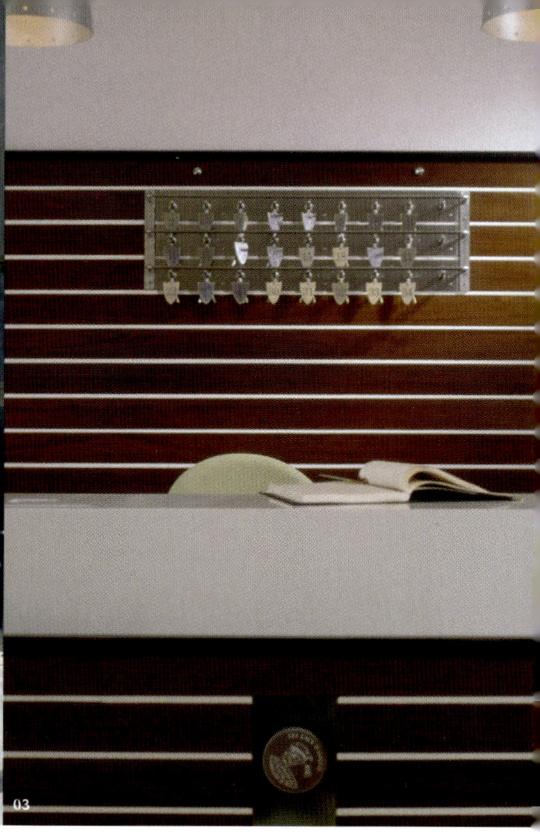 03

04

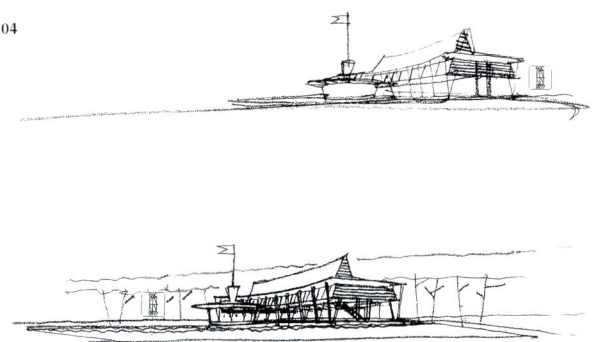

05

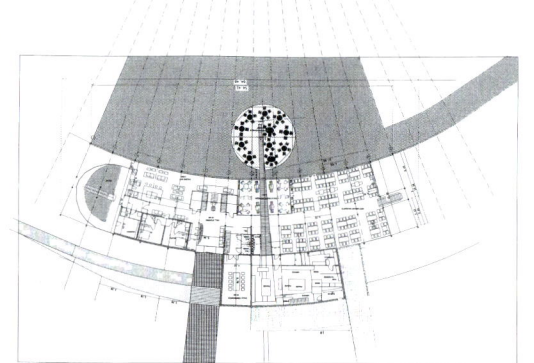

06 07

reflecting the clouds as they pass across the sky. The outline of the building is evocative of Malaysian wooden huts, although the roof has been modelled on local buildings.

Both the furniture in the dining room, which seats 200, and the fittings in the bar were designed and carefully thought-out by the architect Eric Raffy. The reflecting surface of the lake has been brought inside too: a partition made of sand-blasted mirrors has been installed in each room, one side of which functions as a bathroom mirror, the other serving as the headboard on the bed.

The floating rotunda is not only open to guests but welcomes visitors who appreciate regional cuisine. The head chef, Annie Demen, serves up regional dishes with choice wines in this grand setting at regular tasting sessions.

02 | The pre-situated "island restaurant" is the focal point of the building.

03 | Reception.

04 | Design sketches.

05 | Ground Plan.

06 | Swimming baths. The lake is suitable for swimming in the summer.

07 | The three-star house disposes over 24 rooms with a view onto the forest or the lake.

sofitel palm beach | marseille . france

DESIGN: Claire Fatosme, Christian Lefèvre

From a distance, this hotel looks like a luxury ocean liner docked on the famed coastal road of Marseille, the Corniche. The strict, horizontal lines of the hotel blend harmoniously with the rocky coastal cliffs. Sofitel Palm Beach's manager, Domenico Basciano, calls it a "harbour of rest." Among the reasons: Open any window and the first thing you hear is the gentle lapping of waves on the coastline. That's unusual in the bustling port city of Marseilles, with its extraordinary cultural diversity and non-stop night life. A grove of exquisite palm trees gently shields the hotel from daily life in Marseille, and at the same time, gives the entire complex the feel of an oasis on the coast of southern France.

The entrance to the Sofitel Palm Beach is unmistakably French, conveying a love of life and elegance. Through the glass entryway, guests find themselves in a space bridged by a copper roof built around a living palm tree. The foyer affords the first encounter with furniture and fittings by some of the world's best names in design: Zanotta, Moroso and Ingo Mauer, to name a few.

The architectural team of Claire Fatosme and Christian Lefèvre have chosen the twin themes of ocean and nature to guide their design. And variations of each theme are reflected throughout the structure. An underground spring, the Source de Roucas Blanc, is located beneath the building, and the architects have created a sandstone cave for its setting. Nearly all of the hotel's 160 rooms have views of the sea, and windows feature graceful blinds which complement the rhythmic motion of the waves beyond. Rooms are outfitted in dark natural woods and elegant chromed steel, a noble setting for lounges from Zanotta. Lighting comes from fixtures designed by Philippe Starck. Some of the suites include a unique bar extending outside onto a balcony, offering the prospect of intimate cocktails above the shimmering blue waters of the Mediterranean.

Business people and travellers seeking a break from the hectic pace of Marseille have discovered a retreat at the Sofitel Palm Beach. They can be found lounging at the ocean side pool, or relaxing at the hotel's famed spa, the Chateau Berger. Therapies at the spa are part of the exclusive ambiance here, which in many ways is evocative of life on a luxury liner. A stately dinner, certainly fit for the captain of any ship, is also available at the hotel's La Reserve restaurant.

01 | Many of the 160 rooms boast a panorama view over the pool deck and harbour onto the sea.

02 | The architects connected concrete, wood and glass with straight-
   lined architecture. In this area rather an exception.

03 | Lobby.

04 | Wellness landscape in swimming baths.

05 | Additional building in penthouse style. The slats in front of the
   glass front ensure interchanges of light and shade.

06 | Restaurant. The surface of the wall with nautical map.

07 | The spring "Source de Roucas Blanc" was highlighted by the
   planners .

08 | Five room categories are available most of them are deluxe rooms
   with balconies and a view onto the sea.

the st. david's hotel & spa | cardiff . united kingdom

DESIGN: Olga Polizzi

"I believe that every luxury hotel should have its own personality and style that reflects its location and nationality." That is how the chairman, Sir Rocco Forte, succinctly describes the philosophy behind The St. David's Hotel. Those bound for this luxury address will find themselves at the heart of port life in Cardiff, the capital of Wales. The hotel is set right on the waterfront, protected from stormy seas by russet-coloured iron moles, in one of the city's bays. Little fishing boats moored in the harbour basin rock to and fro; one or two chug by. A peaceful picture, and a wholly authentic one.

The hotel complex stands out against this backdrop. It is a large white construction with an eye-catching roof modelled on a bird, whose wings span the whole house. The glass portal over the entrance stretches up to the top storey of the building and reveals the winding balconies

in the interior of the atrium. All lit up at night, with a porter in his appointed place, it resembles a ballroom. And conveys an understated, chic elegance.

The hotel has 132 spacious rooms, including twenty suites, all with private balconies. The wooden planks underfoot and the white railings evoke a feeling of being at sea. The interior design is a mix of young cosmopolitanism and British comfort: lots of upholstered furniture that invites you to just sink down and relax. The bathrooms are a sea of white with blue accents here and there, tastefully punctuated by gleaming steel and mirrors above the washbasin arranged to form a niche.

However, the undisputed highlight of the hotel is its spa. A network of passageways leads through a seemingly endless wellness paradise. The 13

rooms include hydropools, a gymnasium, dance studios and a fitness studio. Numerous Far Eastern or local marine mineral treatments are on offer, including algae wraps or a special therapy involving hot and cold stones that are placed on vital energy points in the body. Guests can indulge themselves in cleansing treatments for the body and mind and round off with dietary advice and a tailor-made personalised training programme.

Imagine, though, if your health check revealed that you couldn't dine in the St. David's Hotel restaurant, the "Tides", where the world-renowned, three-star chef Marco Pierre White calls the shots. But let's be honest, those few extra pounds surely aren't the result of his excellent, light French creations…

01 | The designer Olga Polozzi – sister of the owner Rocco Forte – knows how to combine colours and materials excellently.

02   03

02 | All 132 rooms have a view onto the water, several of them a terrace as well.

03 | Thanks to the room-high glass structures, the rooms are light and offer good views.

04 | Wellness is a focal point in the direction. Adjacent to the swimming hall there is another pool as well as sauna, steam bath, treatment rooms and a detailed selection of fitness activities.

05 | The hotel appears to be an extension on top of a steamer from the other side of the bay.

06 | Atrium with lobby and reception.

04

05 06

babington house | somerset . united kingdom

DESIGN: Simon Morray-Jones

Generally, one would imagine staying in an historic English county to be very romantic. Far removed from the hassle of the city, with long walks, croquet and horse riding, and cosy evenings in front of an open fire. Many country guest houses in the British hinterland play on this image, though rarely hold up to their promises. Stuffy service and a rigorous denial of all things modern-day seem, for many, to pass as eccentricity.

This miserable situation may have been one reason why Soho House Country, sister to the company which runs the renowned Soho House private members club in central London, decided to create a similar, provincial version of their ideal, out of a magnificent property roughly 120 miles south-west of London. The site of Babington House in the county of Somerset was established in the 14th Century as the country seat of the Cheddar family. In the course of time, the property

changed hands and was repeatedly converted, until it was sold in the mid-90s to Soho House Country – a decision which has proven to be a blessing for all concerned. Not only can its own club members holiday here, befitting their club status, but us normal citizens can also find refuge.

With its classical cream-coloured façade, its small chapel and the lush surrounding landscape, the property does indeed fulfil romantic notions of the country idyll. Inside, a completely new interpretation of the "country club" theme is found. Not only has the stuffy atmosphere been replaced in favour of a more open, lively design, but the staff have no hint of formality about them either. Sofas with zebra patterns add to an amusing, kitsch, retro style which greets the guests in the entrance hall. The Manager himself makes a very laid-back impression, and evidently feels very much at ease.

Twenty-seven rooms are distributed over various wings of the house, some of them specially designed with families with children in mind. The rooms in the attic are inspired by a low-key Italian style. The oak floors and the simple wooden furniture appear classic and attractively elementary in the wide, open layout. In contrast, the rooms on the first floor play with popular set pieces, such as a free-standing Victorian bathtub. Every room is equipped with a wide-screen television with DVD player and top-of-the-range stereo system. This bonus, on top of the usual extras, can be explained by the fact that most members of the Soho House club are media and film professionals. In addition to media work areas, the hotel's own 45-seater cinema draws guests in every evening with showings of the latest films as well as old classics.

In view of these extras, it is almost superfluous to mention the first-class service – what else

01 | At one point something completely different: Treatment rooms in the tepee. As an alternative there is the former cowshed with swimming pool or a nomadic jurt.

02 | Babington House is a part of a renowned private club Soho House, with further houses in London and New York, amongst other places.

01 | 02

would one expect? Apart from the pool and tennis courts, the property includes a former cowshed where guests can participate in an extensive wellness programme. A whole range of treatments is offered together with products made on the estate. After indulging oneself here for an afternoon, guests could be enticed by an evening meal in the Log Room or, on mild evenings, on the terrace, to see the day out in al-fresco style. The food at Babington House is honest, rustic, true country fare with plenty of tasty grill and oven-baked specialities, and international influences also putting in a few appearances. The menu changes weekly and is mainly determined by the region's traders who contribute differing ingredients according to the season.

So how does one get to enjoy all these comforts? True, Babington House is, in the first instance, a private club with local residents, within 30 miles, also being eligible to apply for membership. To avoid the influx of people into the region rising dramatically however, the gates of the estate are open to all those looking for something different and a change from the standard romantic country house feel.

03 | The country manor in the county of Somerset is situated in an enormous park with its own small lake.

04 | There are 27 rooms in total, each furnished differently.

05 | The baths were given substantial space.

06 | There are wooden floorboards in most of the rooms. The walls are painted in different colours, mostly in dark tones.

07 | In the so-called Log Room one finds the Restaurant. On the right there is the terrace, which also serves refreshments.

cowley manor | gloucestershire . united kingdom

DESIGN: De Matos, Storgy Ryan

The spa at the rural retreat of Cowley Manor is breathtakingly beautiful. The design has won numerous awards. There is an unmistakable association with Mies van der Rohe's pavilion in far-off Barcelona. Two large windows form the façade; granite posts and walls support the roof. Everything is confined to one level, and the architecture is sleek, light and transparent at the same time. Immersing in the healing waters here is a soothing experience for the mind, as well as the body.

Sharp contrasts between the spa at Cowley Manor and the surrounding countryside are fully intended. It's situated in the ancient, rolling countryside of Gloucestershire, amid extensive wooded greenscapes, not far from Oxford. The manor itself is situated on a small wharf, on one of four small lakes, looking like a stately Victorian waterside palace. Britain's imperial period is reflected in the manor's wood panelled reception area, hallways and restaurant. It is a large property, generously laid out. The grounds are meticulously maintained, providing a wonderful opportunity for elegant afternoon strolls.

Set in this bucolic, historic setting are Cowley Manor's contemporarily designed 30 rooms, plus play areas for children. Walls are a bright green, and playful furniture designed like plush monsters are an invitation to have fun. Rooms vary in size, but all feature light-coloured woods stylishly integrated into their design. Beds, tables and consoles radiate warmth. Colourful furniture completes the fresh, modern ambience. None of this is forced; Cowley Manor is a masterful symbiosis of modern design in a setting of noble heritage.

Part of this unique mix is the manor's outstanding spa. So are the manor's staff, who attend to their guests' every need in the finest tradition of refined British country living.

01 | Glass, wood, cement and wall work shaped into cubic building forms completing the optical contrast programme to the historic manor house.

It is exactly this combination that today attracts London society.

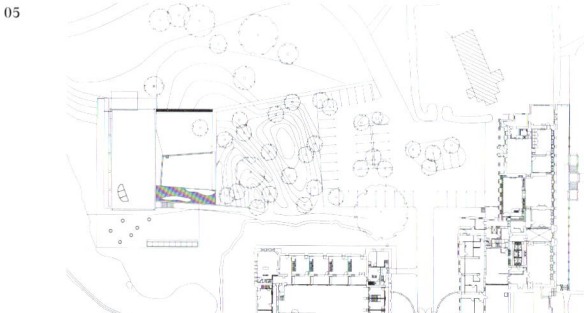

05

02 | Just like in films: Old creaking stairs.

03 | Breakfast rooms and Restaurant.

04 | The baths are laid out with stone plates right up to the ceiling.

05 | Plan of the whole complex. The property is over 22 hectares.

06 | Minimalist bathing hall. The water comes right up to the granite walls.

07 | Section of plan: one of wings of the manor house.

08 | The glass walls of the bathing hall transport a feeling of being in nature.

09 | Architects and owner planned according to the motto: "Generosity before room usage." There are 30 rooms in total.

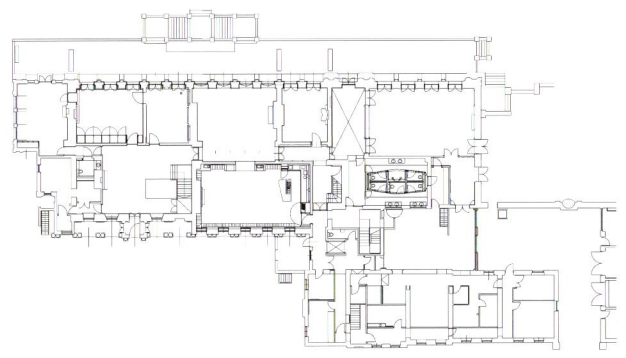

one aldwych | london . united kingdom

D E S I G N :   Charles Mewes & Arthur Davis, Jestico & Whiles, Mary Fox-Linton, Gordon Campbell-Gray

Mention "One Aldwych" to your taxi driver and he'll have no problem taking you there because its just the street number. Upon arrival, however, finding the hotel may prove more challenging. And that's exactly how owner Gordon Capbell Gray envisaged it to be, a hotel of "discreet luxury". Rather than creating an obvious entryway, Campbell has placed an inconspicuous set of double glass doors. In place of a sea of flags, an unobtrusive brass shield tells guests they have arrived. Instead of a lobby with a train-station atmosphere, a bar and lounge with genuine flair are what await vistiors to One Aldwych. This jewel of structural art was erected in 1907 as the headquarters for London' Morning Post by the Anglo-French architectural team of Mewes and Davis, as a duo specializing in hotel design. Though originally planned as an office building, one look at One Aldwych and it's hard to imagine the edifice today as anything but a hotel.

So perhaps it was easy for Campbell Gray to envisage transforming this building into a hotel, because he did so with Mary Fox-Linton, with functional, modern and decorative interior architecture.

The hotel's 105 rooms and suites – two with fitness areas – are maintained in muted pastels, to better bring out the wall paintings and rotating ornamental flower arrangements. Each room boasts impressive city-side views. And their high-tech apparatus – telephone systems, ISDN, and CD players – are all located behind each room's timelessly elegant furnishings. As an extra touch, there are mini-TVs on the edge of each washbasin, perfect for hurried newshounds in the morning. One Aldwych's most unusual offering, however, is its private film screening room with a 35 mm projector. „The Health Club", with sauna, steambath and an 18-meter pool, featuring

underwater music, adds to the surprise. In contrast to the often plain or cluttered wellness environments in other basement floors, here the architecture lends a defining feel-good factor. Lots of glass, aluminium and yellow walls emphasize the blue of the water.

"Axis", One Aldwych's basement restaurant, is quickly becoming a secret find among London's incrowd. It's an architectural mélange of Gotham City and Fritz Lang's Metropolis. Here, as well as in the Mezzanine's "Indigo" bistro and the Lobby Bar, young brokers congregate after the close of nearby city banks. In short: Rarely a hotelier succeeded so well in synthesizing the notions of logic and versatility in a single luxurious private hotel. Surrounded by theatres, boutiques, ethnic quarters and cultural attractions, One Aldwych is perfectly situated at the intersection between London's City and the West End.

01 | Such buildings with pronounced tips
characterised the picture of towns at
the beginning of the 20th century.
Architects, who specialised in hotel
buildings, ironically built the former
publishing house.

02 | The lobby is a popular "after work"
meeting point for the financial world.

03 04

06

03 | Washbasin and bath were designed uniquely. The shower is
implanted in the ground. There is a small TV display next to the
washbasin.

04 | Example of a luxury room. The cupboard front made of fine metal
weave.

05 | Breakfast room and Restaurant "Indigo".

06 | Every floor has a different colour.

07 | Reception at "The Health Club".

08 | 09 On the lower floor of the two-storey wellness area there is
an 18 metre long pool inclusive of underwater music.

07  08

09

sanderson | london . united kingdom

DESIGN: Philippe Starck

The second Ian Schrager hotel to open in London after the whimsical St Martins Lane, Sanderson combines opulence, glamour and style with spirituality and well-being. Another collaboration between Schrager and maestro Philippe Starck, Sanderson is an ironic combination of pure simplicity and extravagance. Much like St Martins Lane, the lobby is an eclectic collection of international furniture and objects that epitomise the fantastical. On the seating side, there is an original Salvador Dali 'Bocca' lips sofa from 1970, a Pierre Paulin 'Tongue' chair and the classic Eero Aarnio 'Hanging Bubble'. A what-would-be conventional Louis XIV style sofa has been stretched out of proportion to 35-foot

length. The reception desk is a frame holding five wide-angle televisions, covered in a sheath of striking green glass. The ever-changing images create flashes of movement. Behind this techno-age counter is a large Louis XV style armoire and the concierge desk is a highly stylised, gold-leaf bureau of the same kind.

Besides the lobby, there are a number of other intriguing public spaces. The Courtyard Garden, designed by legendary landscape architect Philip Hicks in the late 1950s, has been classified as a landmark 'Heritage Garden'. An unexpected oasis, the design is a rather lush version of a Japanese Garden. The original pond and fountain

remain, as does a 40-foot tall tree – a great provider of shade to outdoor loungers and diners.

The eclectic mix of eating and drinking venues are as much about aesthetics as they are about food and drink. The restaurant, 'Spoon+', with its light maple floors, white-clothed tables and ostrich print vinyl banquettes has a menu where diners concoct their own dishes by choosing from a list of ingredients. 'Long Bar' is an 80-foot onyx rectangle that glows from the inside, with custom bar stools. The white-upholstered, silver leaf frames have a photographic image of a woman's eye on the back – somewhat unnerving when all lined up and viewed from behind.

01 | Lobby as a stage for self-portrayal. It shows a cross-section of Philippe Starck's favourite objects.

02 03

The enchanting guest room 'salons' do not pay homage to convention. Beds are grand, silver-leaf sleigh versions that become a playful centrepiece to the room. Linen is white and silver and topped with a fashionable pashmina throw. Attached to the back of the headboard is a desk complete with a white mesh Eames chair. Guests will probably wonder about the lack of artwork, until of course they are lying in bed and are faced with a silver leaf-framed painting on the ceiling above – just one of the ever present quirks. Excerpts from Voltaire's letters grace the custom-made carpets for a touch of the romantic.

Bathrooms are transparent – encased in a clear glass box full of freestanding elements. Not a thing is wall mounted. The spacious Starck tubs are freestanding, as are the stainless steel and opulent orange Venetian glass vanity units. A mirror and a simple pendent light, both suspended from the ceiling above, and appearing to float in space. With the public spaces an overt display of fantasy, the hotel's spa – Agua Bathhouse – is a counterbalance of pure serenity. White, light and ethereal in design, its purpose is to pamper and revive, and a visit to this tranquil oasis must accompany any stay in the hotel.

04  05

06

07

02 | Now often copied: Curtains as room dividers.
The Sanderson offers a total
of 150 rooms.

03 | The easy combination of glass-clear,
straight-lined architecture and playful
Rokoko is considered the typical style
of the designer.

04 | The two-storey Agua Spa is only missing a
pool. View from the upper floor with
treatment rooms in the entrance area.

05 | Changing rooms for men with showers

06 | Quiet zone. Here over six-metre high
curtains serve as room dividers as well.

07 | Changing rooms for women.

seaham hall | seaham . united kingdom

DESIGN: Kappers Architects

Located on England's east coast in the county of Yorkshire, Seaham Hall can be reached in a two-hour drive from Edinburgh, or in three hours by express train from London. The wild countryside of the surrounding area is the perfect backdrop for discovering the region, with nearby Hadrian's Wall or the county's historical capital, York. The estate is sandwiched between two very different natural situations. On the one side is the rugged North Sea coast, on the other an undulating, hilly landscape full of forests and cliffs.

In the midst of this, the opening of oriental baths at Seaham Hall is an unusual attraction for a property so steeped in local tradition. Carved into the rock on which the hotel sits, The Spa, with its teak-clad interior, offers an Asian ambience with spectacular waterworks. An underground path directly connects the grotto housing the spa complex to the hotel, and a watercourse accompanies guests on their way through the darkness, setting the mood for the refreshing and decadent pleasures to come. A 20-metre-long swimming pool, hot springs and a waterfall are merely a few of the highlights of the magical cavern, which also offers Thai massage, various cosmetic treatments and even an Asian cookery course.

Nineteen individually designed rooms, fully air-conditioned and luxuriously furnished, await the hotel guest. As standard expect a Bang & Olufsen TV, in addition to internet access, a direct dial telephone and fax machine. Furthermore, guests are able to hold video conferences or multimedia presentations in four private conference rooms, which can accomodate 120 visitors altogether. This arrangement especially recommends Seaham Hall to companies wishing to hold their congresses and seminars in a relaxed environment.

Guests can choose between more or less extravagant rooms according to their taste and budget. The categories range from single suite to penthouse. All the rooms are equipped with a large bathroom with double washbasin and a bathtub big enough for two. Optically, the rooms are strong, but tasteful and slightly reminiscent of a neo-Biedermeier style. With their matching colours and sensitively coordinated material, however, they are soothing on the eye and appear more modern than antiquated. The public areas also serve as an art gallery for paintings and sculptures, the main focus being on regional and Norwegian artists. Especially worth seeing is the ceiling of the atrium-type lobby designed by the stained-glass artist, Bridget Jones.

01 | View from one of the 18 suites.

02 | The country manor, located on the English east coast, was built in 1792 and opened as a luxurious conference and wellness resort in 2001.

03 | 04

03 | Decorative accessories and art objects were selected carefully.

04 | Conference room with view onto surrounding park.

05 | Five different and completely luxurious types of rooms can be
opted for, crowned by the penthouse suite.

06 | Although they are anything but minimalist, the rooms seem calm
and harmonious.

07 | All the rooms have a two-person bathtub.

05    06

07

atoll | helgoland . germany

DESIGN: Alison Brooks

What is it that makes this tiny part of the earth so special? Helgoland: 3.7 km² of sandstone rock, jutting out of the North Sea, swept over by the warming Gulf Stream winds. The sunniest location in Germany, miles from big town hectic and industrial noise. The oxygen-rich, unpolluted air and a mild climate make this high-sea-island a natural oasis, more popular today than ever before.

The island has been a preferred destination for more than a century. At the start of the 1900's, its reputation as an exclusive bathing resort with an elegant atmosphere made it a suitable place for aristocracy to meet with financiers and

industrialists. This period of chic and Savoir Vivre came to an abrupt end, however, in the form of Kaiser Wilhelm II's ambitious plans to turn the island into a fortress and military base. This was the beginning of the end, dragging in a painful epoch that lasted for decades and very nearly culminated in the total destruction of the island at the end of World War II. Luckily, Helgoland was spared this disaster, but many years and immeasurable tenacity were required to re-build the rock, and create an environment to entice the tourists back.

The owner of the Atoll, which opened its doors in 1999, followed a marketing concept that made

the most of the island's extraordinary qualities. His vision of a contemporary hotel was based on four pillars: wellness, business, gastronomy and design. He offers his guests the best possible work facilities for meetings and conferences, complemented by outstanding food and drink, and an atmosphere of fantasy and originality, created through the hotel's extravagant design. These three combining elements are rounded off with the wellness and fitness area that allows guests to optimize the free-time they have earned after a hard day's work.

The hotel comprises of 43 double rooms, 7 suites and one apartment, some of them classically

01 | Business, design, gastronomy and wellness combine the hotel on the 3.7 square kilometre north sea island.

02  03

04

05　06

designed, some futuristic. The London-based designer, Alison Brooks, constructed the interiors with a dedication to detail, where the language of forms flows through the entire concept in the use of rounded edges, sine curves and waves. Her guiding principle was to create a stage, free of any of the coventional furniture or trappings normally found in a standard luxury hotel room. At Hotel Atoll she has succeeded, fashioning living spaccs where objects play different roles, changing and evolving through a story, almost as characters. A table folds out into shelving that, in turn, snakes along the wall before mutating into a seating bench. The use of glass, metal, mirrors and special materials that are normally more at

home at NASA than in the North Sea has made the hotel into an unmistakable, distinctive place. Wierd and sometimes comic, unexpected sight lines and views mark Alison Brooks' style.

No less remarkable are the four locations around the hotel taking care of hunger and thirst. Seafood is the keyword and, as one would expect from the briny location, the chef covers the whole spectrum. Whether in the restaurant, bistro or Sansi-Bar, even the most jaded of palates will find a taste to remember.

02 | 43 double rooms can be found as well as seven suites and one apartment.

03 | Alison Brooks' strong colours and flowing forms are seen most clearly in the public areas.

04 | Compared to that the swimming pool, sauna, steam bath as well as treatment and quiet rooms are very calm and light.

05 | Even when wind and weather make the view dull, the light and transparent architecture is at the forefront.

06 | The bar also belongs to the eye-catchers.

grand hyatt | berlin . germany

DESIGN: Jose Rafael Moneo, Hannes Wettstein

In the heart of the business and entertainment district around Potsdamer Platz lies one of the city's most luxurious hotels. Flanked by cinemas, shops and shopping arcades, it gives guests a taste of the new buzz in the capital. The Grand Hyatt Berlin is on the new Marlene Dietrich Platz, in the midst of Daimler City that was modelled on cities in old Europe, although with fewer and taller buildings.

This is not the first example of the new Berlin "Classicism" flirting with classical elements: regularly-spaced windows in a punctuated façade, a plinth-storey-roof system of building and a subtly imposing ambience. Jose Rafael Moneo, the world-renowned Spanish architect, was responsible for designing the building; the Swiss Hannes Wettstein for its interior. The image created is not one of Mediterranean exuberance and frivolity, which would not reflect

the character of Berlin, but rather one of elegance fashioned in stone. The planners have successfully avoided any kind of clumsy and exaggerated solemnity. In the hotel's 326 rooms and 16 suites, including two very luxurious presidential suits (161 and 210 m^2 respectively), dark wood contrasts with cream-coloured walls, traditional patterns on curtains and covers juxtapose with strong colours, and warm, cosy lighting in the living spaces is set against more functional light in the workspaces.

Although, or perhaps because the hotel primarily lives by its business customers, the rooms have extravagantly large bathrooms measuring nearly 20 m^2. Each is fitted with a shower, a free-standing bath tub and a wardrobe, and provides lots of space for guests to pamper themselves. The fact that the marble floors and even the mirrors are heated seems almost a minor consideration. Along with these more private

havens, the Grand Hyatt houses arguably the city's most beautiful wellness areas on its spacious, glass-fronted top floor. Guests can sweat it out in the sauna, take a swim, have a massage or beauty treatment – and enjoy a fantastic view into the bargain.

The hotel's gastronomic delights are also appreciated by the locals. The most popular of its restaurants, the "Vox", is named after the first German radio station which broadcast music and entertainment from the very same spot. You can even watch the chefs preparing sushi, frying meals in woks and cooking delicacies in a wood-burning oven in the showcase kitchen. The "Tizian" is oriented more to Italian cuisine and the "Dietrich" serves small, American-style dishes. All three restaurants offer healthy meals made from the freshest ingredients to complement the hotel's wellness concept.

01 | An eye-catcher in the Lobby is the
sharp crystal-type glass object.
In different colours according to
daylight.

02  03

04

02 | DaimlerChrysler is the investor in the house, which also wanted to create an architectural milestone.

03 | Always praised: the Spa & Fitness Club Olympus with view over Potsdam Square. The swimming pool measures 15.5 x 5 metres.

04 | Sauna, steam bath and rain shower.

05 | Changing rooms.

06 | 342 rooms are a lot. Generosity has not been spared.

grand hotel bellevue | gstaad . switzerland

DESIGN: Eric Reichenbach, Gottfried Hauswirt, Rosemarie Horn

It's the stuff that great hotel stories are made of: Take one successful entrepreneur, a jet-setter, who spends more nights in hotel rooms around the world than at home. He registers what works and sees, or rather knows from experience, what doesn't. And then grabs an opportunity that comes his way back home. In our particular example a hotel was put up for sale in the holiday resort of Gstaad in the Saanenland area of Switzerland. But it wasn't just any old hotel, more a palatial villa that dreams are made of. Thomas Straumann made his childhood dream come true and bought the estate in the middle of an idyllic park on the outskirts of the village. He then pumped a great deal of money into renovating it, although the exact sum is a closely kept secret. That it was more than peanuts is obvious from the fact that the newly constructed wellness area covers a total of 2,500m^2 – and the hotel only has 35 rooms.

The complex, originally built in 1913, was completely gutted and two towers added to the late-Art Nouveau façade. Those elements that seemed worth saving, like the wonderful Art Nouveau stairwell, with its stone steps and wrought iron banister, the stucco plaster on the walls and ceilings, parquet flooring and crystal chandeliers, were lovingly restored. Careful study of the names given to the restaurants and the jazz bar – Prado, Coelho and 911 – reveals a great deal about the owner's predilections and hobbies. His personal taste is reflected in the rooms, too, which are partly inspired by the Far East, and furnished with classic 1940s, such as Corbusier leather seats and sofas or have a Philippe Starck interior, like for example the Etoiles Suite.

Reclining in the free-standing bathtub, guests can admire the stars through a glass dome.

Straumann has even found a pithy catch phrase to sum up his philosophy: "Wellness dedicated to opening closed minds". This includes exquisite cuisine and intellectual fodder in the form of literature and films. The piece de resistance is the new spa: a transparent chalet built out of wood and glass which lights up like a torch at night. Clear lines and forms are characteristic of the architecture and only a small number of select materials, such as the anthracite-coloured stone from China or the bleached larch-wood flooring, have been used. Guests can relax in the Asian pool with its jacuzzi and on tatami beds whilst being massaged, in the thalasso pool, in steam baths in the Roman-Ottoman tradition, saunas, indoor pools and in a rest room that looks out across the newly designed Japanese Garden in the sheltered atrium courtyard.

01 | The Japanese Suite measures 60 m² and is on the fifth floor underneath the roof. The partition is made of rice paper and wenge sprouts

02 | Guests can enjoy their favourite films on a high-tech screen in total privacy.

03 | A wine tasting session in the wine cellar, where guests are surrounded by racks laden with fine wines.

04 | Wine crates from the bar in the cellar.

05 | Asian bathroom with a free-standing bathtub "New Haven".

06 | The gourmet restaurant "Prado" with pictures by contemporary Swiss artists.

07 | The bathroom in the Etoiles Suite where stars twinkle through the glass dome at night

02

03

04 05

06 07

lenkerhof alpine resort | lenk . switzerland

DESIGN: Rolf Balmer

How do you go about creating the most youthful five-star hotel in Switzerland? That was the challenge that the Lenkerhof set itself. The hotel is situated at the southern tip of the Bernese Oberland, the most beautiful place in the world – at least that's what the people of Lenk will tell you. The answer was actually quite obvious, at least to designer Rolf Balmer: "By being playful". And playful is the term best applied to the interior design of the Lenkerhof since it reopened after months of renovation. Humorous allusions are dotted all over the place: you can relax on a cow hide armchair in front of the fireplace; the chandeliers in the hall, which are 3 meters tall and made from rusty iron, are crowned by a corps de ballet of cows – alluding to the Berner Oberland where grazing cows are an integral part of the landscape.

In the à la carte restaurant "Oh de Vie", which

serves light, imaginative cuisine, guests are seated on angel chairs. The idea is that they should be carried away by the culinary delights. Above the beds in the guest rooms are quotations taken from 150-year-old tourist brochures. Flying in the face of the current trend, old and new blend together so that you cannot tell where the old ends and the new begins. The expansive hall is a sea of fur, velvet and soft pile. One hundred and thirty candles are lit here every night, providing a dazzlingly romantic production in which the main role is played by Catellani and Smith lights. The same could be said of the "fil de fer" lights above the bar: balls made from silver wire reflect the light a thousand fold.

The 2000 m² spa has a pristine clarity throughout; blue and red symbolise warmth and cold. The name "7 Sources" is a pun on the seven sources from which the local river, the Simme, rises.

Natural stone originating from the Simme was also used to build the spa. A dip in the Ice Grotto is a pleasant way to cool down after a sauna.

The Crystal Bath offers a holistic musical experience: the tub serves as the resonating body, the sound is transported via the water and is taken up by your body; the water also changes colour depending on the intensity and style of the music. Each treatment cubicle offers a wonderful view of the mountain range, especially the striking Wildstrubel Mountain. The spa aspires to be extrovert, awakening the senses. The key to its philosophy of well-being is activity and sports. "We might be mentally tired today, but the best was to regenerate is to be physically active," says the Director, Philippe Frutiger. He will even accompany his guests in person on their skiing tours, mountain walks and yacht cruises on the Thuner lake.

01 | Outdoor pool with a view of the Wildstrubel Mountain which marks the divide between the northern and southern Alps

02 | The rooms are furnished in the classic style prevalent in the old
hotel.

03 | An oversized chaise longue greets guests in the entrance hall.

04 | The original style of the Smokers' Lounge and its humidor has
been retained.

05  06

07

05 | Red and blue symbolise warmth and cold in the spa area. The
natural stone was taken from the Simme River.

06 | Indoor pool with a view of the Bernese Oberland.

07 | The winter garden provides sheltered access to the outdoor pool,
which has been fed by a sulphur source for over 350 years – the
strongest in Europe.

parkhotel weggis | weggis . switzerland

DESIGN: Vincenz Erni, Vadian Metting van Rijn, Müller + Partner AG, Jörg Richli, Pius Notter

Weggis is only a 30-minute steamboat ride across the Vierwaldstätter See from Lucerne. It was the British who discovered this little village on the west bank of the lake, under whose particularly mild microclimate palms and olive trees thrive, at the beginning of the last century. Illustrious guests flocked to what was dubbed the "Riviera in the heart of Switzerland": Mark Twain, Serge Rachmaninoff and Alexandre Dumas have been honoured by having the most beautiful suites in the Park Hotel named after them.

This Art Nouveau jewel lies on the banks of the lake in the middle of a 22,000 m^2 park, with a magnificent view of the chain of mountains that form central Switzerland. It has been updated for the third millennium with a great deal of commitment and at huge financial expense. The entrepreneur Martin Denz, who spent his childhood holidays in Weggis, has turned his own personal vision of a perfect holiday hotel into reality. This is a five-star hotel which is grand but never ostentatious: a place where "you can do everything, but don't have to do anything". The original beauty of the main building's Art Nouveau façade and that of the "Schlössli" have been restored; the interior has been designed to give it a contemporary feel, with bright colours. Paintings by the Swiss artist Susi Kramer make it a cheerful place. High-grade materials, oiled American cherry-wood and matt grey slate for flooring, stainless steel for the door frames have been combined with designer furniture and lighting, seating by Molteni, Flexform, Geravasoni and Cassina, lighting by Philippe Starck and Murano. The bathroom in the Mark Twain Suite comes replete with centuries-old Cotto tiles, a free-standing bath tub, a view of the surrounding terrace and untreated mahogany floorboards, and was awarded as most beautiful bathroom in Switzerland. In quiet understatement the gourmet restaurant, which serves light, Mediterranean cuisine, has been called the "Annex". The colours were chosen to match those of the Versace crockery – not the other way round.

01 | "From Lucerne to Weggis ..." on the steamboat trip over the Vierwaldstätter Lake to the park hotel – one is reminded of the Swiss

yodelling song – an authentic soundtrack to the panorama of the Alps.

02 | 03

The newly built Aquarius Hall, which can be booked for events, has been given a truly avant-garde design: a rectangular, six-metre high cube with a double shell of etched glass is embedded in a Japanese garden laid out by the Swiss Pius Notter, who is also highly respected in Japan. An LED light display featuring 90,000 lamps set in the façade illuminates the entire hall in an array of colours.

The spa also has innovative features. Six spa cottages, each 70 m^2 in size, are set around the Japanese garden and can be booked individually. They offer wellness in a very private atmosphere. Treatments use only natural products. Guests can enjoy a flower bath, break out in a moderate sweat in the tepidarium or simply relax on a water bed and listen to the sounds emanating from the Dolby system – all in a Far Eastern ambience.

02 | Lobby of the Aquarius Hall with sofas by Cassina.

03 | Strawberry-red has been used in the Rachmaninoff Suite under the white beams.

04 | The carefully restored Art Nouveau staircase.

05 | View from the Japanese garden to the futuristic Aquarius Hall.

06 | Cross-section through the complex. The block behind the old building houses the conference centre. Part of the spa has been built into the hills. Daylight pours into the dome.

07 | Foyer. Passage to the Spa cottages and the Japanese garden.

08 | Dream away the hours under the dome that provides the Flower
Bath with daylight

09 | Your own private steam bath.

10 | Mahogany and water are the main elements.

11 | The cottages have been named after precious stones.

12 | View of the dome through which daylight floods in.

therme vals | vals . switzerland

DESIGN: Peter Zumthor

Connoisseurs of good mineral water have known about this little Swiss village high up in the mountains for a long time, as it is the source of the fine Vals water. However, since 1998 the name has also been a catchword in architectural circles. The newly constructed spa set in the cliffs, along with its affiliated hotel, has caused quite a stir since opening and has attracted many an enthusiast of architecture. Peter Zumthor designed and built the spa using local Vals quartzite rock. His premise was that "what I design will be part of this place, part of its surroundings". He has created a building that is like a complex piece of music composed of ever changing variations on the four main themes of stone, water, light and air.

Energy flows through the spaces in the building and new perspectives and atmospheres are revealed with every step you take. The omnipresent grey rock, which also dominates on account of its dark colour, gives the edifice an archaic character. A total of 60,000 slates have been piled up in precisely arranged layers to form the massive bearing walls.

Although only a single type of stone has been used in the entire complex, it is never boring or redundant. Great variety has been achieved merely by treating the surface in different ways, ranging from roughly hewing to grinding to polishing it. A virtuoso performance acted out by light, shadow and reflections on the surface of the water unfolds during the course of the day. The ensemble is at its most impressive in the early hours of the morning before the endless streams of day-tourists have completed their arduous pilgrimage to get here. Now hotel guests have the rare privilege of being able to walk in these hallowed surroundings, virtually alone and undisturbed, watching the first rays of sun, or taking a glass of the sought-after Swiss mineral water that springs from the rock.

The baths range in size, lighting effect and temperature, from the extremely cold ice pool and the flower pool to the steam bath. Special facilities such as the music rock and the grotto pool complement the range of offers. To boost your well-being you can indulge in an exfoliating massage, a body wrap with algae or herbs and flowers or meditative breathing, to name but a few of the treatments available.

The adjoining hotel, a high-rise built in the 1960s, had a forced cosy-Alpine feeling until Annalisa Zumthor and the TV journalist Claudia Knapp took it in hand. They have brought a breath of fresh air that has banished the old and musty atmosphere. Although neither of the two new managers ever trained in hotel management,

01 | The Thermal in the furthest point of the Val Lumnezia has developed into one the most attractive points for those interested in architecture.

06 | 07

they have imagination, an understanding of art as well as the determination to raise the hotel to the same standard as the spa. The hotel has a total of 46 four-star rooms and 80 three-star rooms. Peter Zumthor has already redesigned 19 of them in his own unique style: white cement floors, blue-black carpets, specially designed bedside tables and linen sheets. Instead of the obligatory TV, the rooms each have a CD player with select discs. The prevailing motto is one of healthy enjoyment: after a massage or beauty treatment and a freshly mixed fruit/vegetable juice, guests can lie down, listen to some music and take in the beautiful Val Lumnezia. It's like a dream or a childhood fantasy.

02 | The interplay between material and light is unique.

03 | Solid rock in nature and refined in the building work.

04 | Quiet zone. No question about it – the architecture polarises: either one is enthusiastic about the purity or disgusted by it.

05 | The water reaches right up into the walls.

06 | Quiet zone with view onto the scenery of the Alps.

07 | Annalisa Zumthor and Claudia Knapp manage the attached hotel from the 60s in which Peter Zumthor renovated the total of 126 rooms gradually.

rickatschwende | dornbirn . austria

DESIGN: Thomas Hämmerle, Marika Marte Enea

Rickatschwende, which lies 850 metres above Dornbirn in the area where Austria, Switzerland and Germany meet, attracts those who want to take time out to think about their lifestyle, perhaps radically change their diet or do something for their body and their soul away from the opulence of everyday life, far from the frantic rat race and overstimulation of the senses. F.X. Mayr regeneration therapies, anti-ageing strategies, anti-stress and re-energising programmes are among the treatments in which the Rickschwende Health Centre specialises. The goal is to help guests cut back, to pause, to find inner peace.

All this is reflected in the immediate environment: architecture has been used to express the philosophy that underlies the concept. The architect Thomas Hämmerle, son of the family that owns the hotel, chose clear perspectives and linear design as his form of expression.

And less of everything; simplicity, but top quality. This concept is also revealed in the understated design of the rooms and suites. The combination of materials, for example walnut, or the oak parquet flooring, creates the right atmosphere, along with sparing use of top-quality furniture by names such as Flexform and Wittmann.

Originally Rickatschwende was a mountain inn much frequented by walkers who enjoyed the view of the Rhine valley, the southern side of Lake Constance and the Swiss Alpine peaks from its sunny plateau. In 2000 the wellness area was renovated and an extension constructed out of glass and wood was added to the former inn, which retained its typical brown wood shingles and shutters. The indoor pool, bio-sauna with colour light therapy, steam bath, shiatsu pool, stone Kneipp circuit and relaxation room that is flooded with light promise active relaxation.

Numerous therapies are on offer and guests can have a programme specially put together to suit their individual needs. The therapies aim to provide direction in a life in which stress and relaxation, action and inaction, movement and resting should be in equal balance. Personal coaching is available to assist guests activate their own resources and to deal responsibly with their own body.

The large sun terrace was designed by the Swiss horticulturists Enea, whose goal is to seek in our gardens the last traces of a paradise lost. In their clear concept and by restricting themselves to only a few types of plants such as lavender and beech they have followed the linear design principles applied to the hotel itself.

01 | The health centre features a Kneipp circuit
with stone seats for a warm footbath.

02 | Less is more: a suite in the new building.

03 | Nature takes centre stage in the glass-fronted relaxation room.

04 | The design principle here is one of clear lines.

05 | The sun terrace offers guests a magnificent view of the Rhine
valley and the southern side of Lake Constance.

03 04

05

post hotel | bezau . austria

DESIGN: Oskar Leo Kaufmann, Johannes Kaufmann

With its solid, wood-shingled houses, the Bregenz Forest feels a bit sleepy, but in the nicest possible way. Far away from the busy mountain regions, the Bödele is where people come to spend a few quiet days. But remember that looks can be deceiving: this is the home of the avant-garde, at least in terms of its buildings. The Austrian region of Vorarlberg is an eldorado for anyone interested in contemporary architecture, and it is also the home of one young proponent of this school, Oskar Leo Kaufmann. His modular townhouse – conceived for the British lifestyle magazine "Wallpaper" and later presented at the Milan Furniture Fair – caused a sensation.

The hotel which his sister Susanne now runs has been in the family for five generations. Kaufmann planned an extension with avant-garde rooms to consciously contrast with the bourgeois feel of the main building. Since the work had to be completed quickly, a creative solution was required: the extension was constructed in only six weeks using a system of wooden boxes – modules completely prefabricated in carpenters' workshops. Kaufmann's transparent architecture has created space for the surrounding environment: the soft mountain pastures of the Bödele, the rich meadows right on the hotel's doorstep. The straight furniture, made from American cherry-wood, was specially designed for the Hotel Post and is pleasingly understated.

Because demand far outstrips supply, an additional 11 classic rooms have been added to the main building which all bear the distinctive Kaufmann hallmark. Painted a light cream colour and furnished with classic B&B, Edra and Alias, they have a modern, elegant feel. The typically 1970s flowered tiles that have been retained in the bathrooms add a quirky touch. However, the highlight of the wellness area replete with saunas and an indoor pool built in the 1980s, is a black-tiled whirlpool out on a first-floor terrace. The wonderful view is doubly pleasing when you can enjoy it relaxing in luxuriously warm, bubbly water. A professional beauty salon and Chinese medical treatments round off the range of offers.

01 | Wooden boxes create a feeling of transparency.

02 | 03

02 | The baths are simply separated from the sleeping areas by a glass membrane.

03 | Transparency as a principle of design: an avant-garde room.

04 | Creamy white dominates the classic rooms.

05 | 06 New extension in isometry.

07 | Whirlpool with a view.

08 | A chrome steel staircase leads to the quiet room.

04

05

06

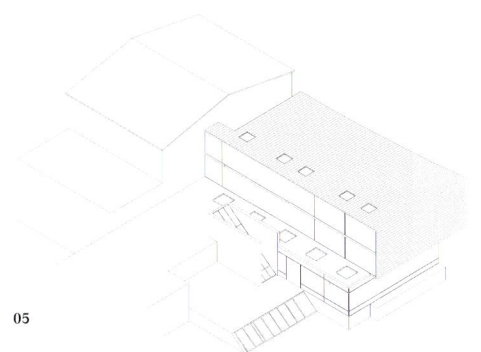

07 08

design hotel madlein | ischgl . austria

DESIGN: Mescherowsky Architekten BDA

It is in Ischgl of all places, the "Alpine Ibiza" as it has been dubbed, with its wild après-ski nightlife, that the hotelier Günther Aloys has created a haven of contemplative peace. In fact, this is the same man the Tyrolean skiing resort has to thank for mega-events on snow-capped peaks featuring the likes of Elton John, Bon Jovi and Michael Jackson. Günther Aloys, whom people tend to either love or hate, says he is only being consistent: "It's precisely in this kind of hell that you have to give people a place where they can find peace." The architecture of his hotel extension, inspired by Far Eastern Zen philosophy and given concrete form by the Mescherowsky's, plays with contrasts. The rooms are characterised by a clear language of forms.

Stone, wood, glass. The materials have been chosen to sensitise perceptions to the opposites of hot and cold, rough and smooth. The entire flooring consists of 1m-by-1m granite slates which were sand-blasted to give them a velvety surface that is pleasant underfoot. The minimalist design in the rooms dispenses with anything that is purely decorative. The mauve-coloured, stained wood harmonises perfectly with the grey, soft seats and the matt, white-washed oak parquet floor. Only a glass wall separates the spacious Phillipe Starck bathroom from the bedroom in each room. The balconies have wooden decking and panes of glass instead of railings so that nothing blocks the dramatic vista of the mountain top.

Gigantic glass fronts in the wellness area create a fluid transition from the inside of the hotel to the mountain world outside, giving you a clear view of the Zen garden that was designed according to an exact plan. Nothing has been left to chance. Teak planks have been inlaid in the slate ground to form pathways linking various areas – indoor pool, beauty, saunas – and continuing the theme of opposites created by lines and areas. The Fire Room is a meditative place – quite the prehistoric experience. After a stressful day on the pistes you can sit round the camp fire that is in a sunken area in the floor. The wellness area exudes a feeling of protection and warmth to give you time to reflect on yourself. "That is why the environment should not be emotionally charged – nothing should deflect the senses," says Mescherowsky.

01 | Quiet zone in the wellness landscape.

02 | Plain furnishing and pastel colours
dominate most of the 80 rooms.

03 | Example of how atmosphere can be
created using small details.

04 | Glass walls create light flooded rooms
with a panorama of the Alps.

05 | The minimalist staging is a pendant to
the Alps yodelling style.

02

06 | 07

06 | 08 Wellness centre with a view onto the swimming pool.

08 | Altar-style wall design with calming effect.

09 | 10 Asian Zen garden meets Austrian Alps.

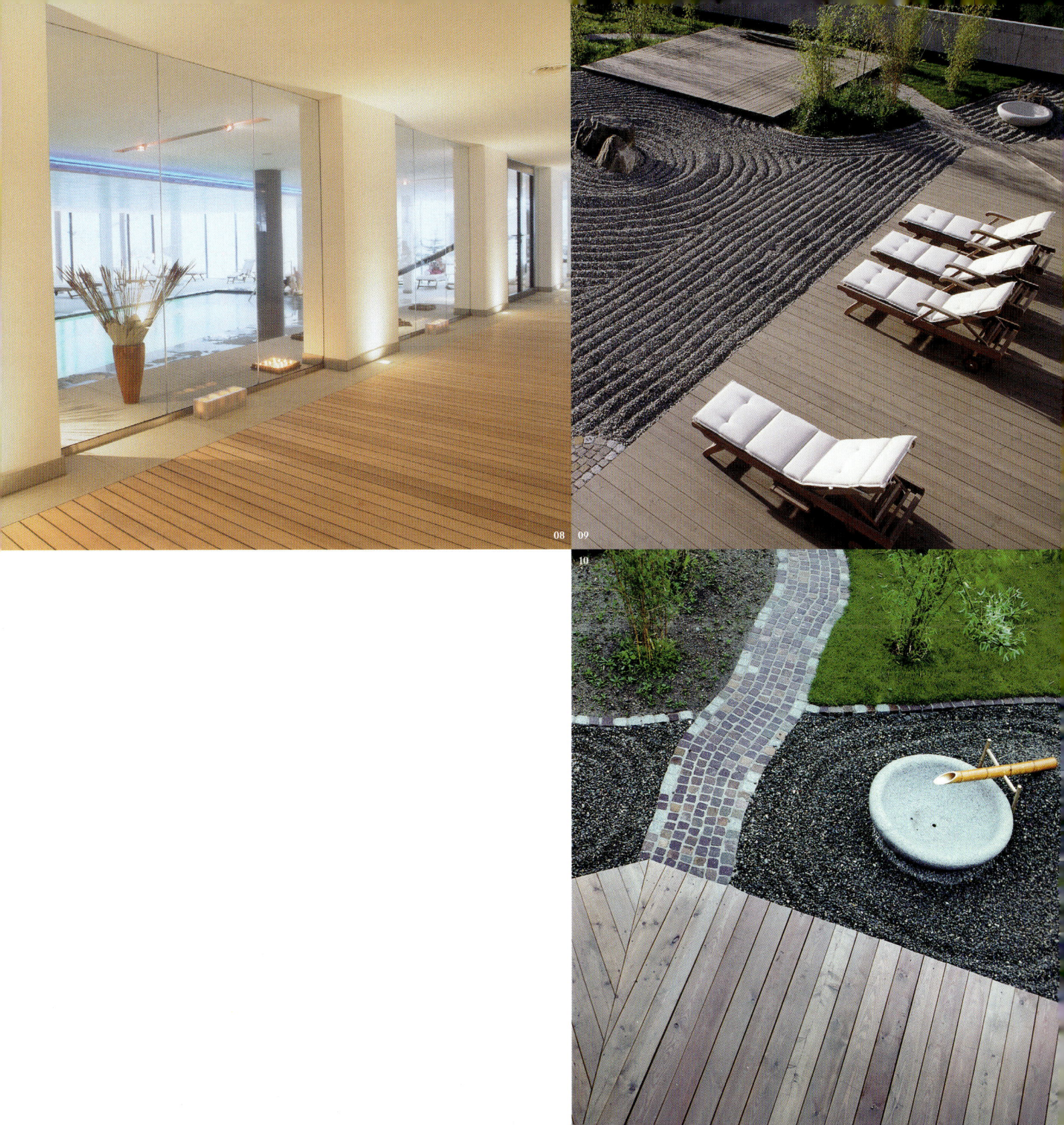

10

schwarzer adler | kitzbühel . austria

DESIGN: Wolfgang Pöschl

At first glance the Schwarzer Adler is exactly what you would expect from a hotel in Kitzbühel, the skiing resort favoured by celebrities. Built in 1986 in the typical Tyrolean style, it features traditional wall decorations, painted window sills, oriels and window shutters. However, behind the sedate façade lies one of the most innovative spas in Austria. When a new generation of the old Kitzbühel hotel dynasty took over, it brought with it a raft of new ideas. Christian Harich, the son with an interest in contemporary architecture, took his inspiration from his many visits to Asia. He finally found a partner in Wolfgang Pöschl, the architect responsible for the original and modern Aparthotel Anton in St. Anton, and together they realised their vision of a spa which is worlds away from the seemingly ubiquitous Alpine baroque style.

As the hotel is situated in the centre of Kitzbühel, land was in short supply and this posed a particular challenge. So they had to dig deep: the swimming pool and saunas are below street level and light wells have been strategically placed to allow daylight to flood in. The Black Spa is a haven of peace stretching over three different levels and a total area of 1100 m^2 that offers visitors the opportunity to withdraw from everyday life and lose themselves in contemplation. The hip name is a pun on the name of the hotel ("Black Eagle") as well as on the black stone that was used to build it. The concept involved paring down everything to a minimum, which also meant being restricted to using only a few materials: stone, concrete, oak, mosaic tiles. Guests can relax in the sauna area which includes a laconium and spa mud bath or in the pool with underwater music. Special features include the Meditation Room with its open fireplace around which waterbeds are grouped, each fitted with its own stereo. The fitness studio, with its modern training machines, provides what must be the most fantastic view of the Kitzbühel Horn from behind the panoramic glass front.

The unusual concept on which the spa is based is also reflected in the range of treatments available there: "Energy balance", a synchronous massage using a soft powder made from fragrant plants and woods, aims to balance mind and body. "Aqua motion", promising undreamt of, deep relaxation, is a special massage technique in which trained therapists mobilise the body's energy as you glide through the warm water listening to sounds coming from the underwater stereo system.

The rooms and suites in the hotel stand in stark contrast to the innovative spa, since they have largely retained the rather traditional country house style. But there are plans to extend the innovative new style into this area too.

01 | 02

01 | A profile glass wall protects the quiet
     area.

02 | Change in perspective: View into the
     quiet area.

03

03 | Clear lines, only a few materials: the pool with an underwater music system.

04 | Cave feeling: In the Meditation room water beds are grouped around a fire area.

05 | Warming bench with 'Kneipp' sinks.

06 | Hall from the reception to the winter garden.

vigilius mountain resort | lana . italy

DESIGN: Matteo Thun, Christina von Berg, Studio Thun

The wood in the dramatic sunscreen on the façade of the Vigilius Mountain Resort has been recycled from an old barn. It had been weathering the harsh climate of southern Tyrol for generations. This natural curing process has refined the material to its purist characteristic: strength. It's a concept which is reflected in many aspects of this unusual mountain retreat.

Milan-based architect Matteo Thun had a single goal in mind when he designed this hotel. He wanted to create a facility to wean urban dwellers from the polluted air and hectic pace of the modern metropolis. The structure he built was supposed be a source of clarity of thought and purity of intent for the weary. In the same way weathered wood was rejuvenated by incorporating it into a sleek and functional façade, the resort is redefining the surrounding area as a place of rest, healing, and Zen-like contemplation.

There is no road leading to this resort. Sounds of civilisation are trapped in the valley below. One of Europe's oldest cable lifts, in fact, is still used to reach the resort. Stepping off the lift directly into the lobby, guests are absorbed into the sleek architecture of sharp edges and clear lines. Thun has chosen to quote the basic structure of buildings typical for this rugged landscape, which blend so harmoniously into the life force of the mountains. The grass-covered roof of the resort is an ecological as well as aesthetic manifestation of this principal of co-existence.

Rich woods and textiles in natural tones promote a mood of contemplation throughout the interior. Windows stretching from floor to ceiling provide unimpeded views of the outdoor surroundings, reinforcing the feeling of oneness with nature. Private quarters feature living and bath areas separated by heated stone sculptures. And the

spa at the Vigilius Mountain Resort is a temple for the spirit and body. It features saunas, steam baths, whirlpools, as well as Sebastian Kneipp water therapy. These treatments are highly effective in stimulating the circulatory system until it has rediscovered its natural balance.

But the mountains also beckon guests to venture out onto their paths to experience the wonders of nature first hand. Personal trainers function as guides, accompanying guests on vigorous mountain biking treks or rock climbing adventures. Back at the resort, massages are available to soothe tired muscles and joints. Spiritual and physical energy must be expended to clear the body of the negative influences of modern urban life. This is the right place for those willing to make the investment.

01 | Mattheo Thun's first Alps resort promises a new type of combination between architecture, the mountain world and wellness.

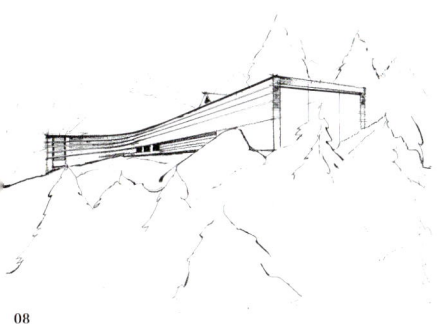

08

09

02 - 05 | Consequent – not to say radical – Thun quotes the building style of the traditional barn.

06 | All 31 rooms and 8 suites have room-high window surfaces and wooden terraces.

07 | Heated walls made of African soil separate living and sleeping areas from the baths.

08 | 09 Sketches visualising the architecture blending harmoniously into the landscape.

aleph | rome . italy

DESIGN: Adam Tihany

The meek may well inherit the earth, but they won't be getting their hands on the Aleph. Exclusive in the truest sense, the hotel is a celebration of luxury, decadence and idle pleasures, formed around the eternal themes of heaven and hell. Designer, Adam Tihany, hasn't pulled back from taking risks in the Via di San Basilio, where this robust former bank building has kept its old face, but been given a devilish new soul.

With the ground floor shrouded in mysterious tones of the underworld, nirvana in the basement and guests sleeping in purgatory, the Aleph is flamboyant, provocative, eclectic. Climbing the

hotel steps, guarded by dozing, caged marble lions, guests are immediately bathed in suggestive crimson light, coloured through the glass entrance doors. Sharp-suited staff welcome with a deferential nod and smile, but the over-sized samurai warriors that dominate the twilight front lobby don't look quite so approachable, towering left and right. One representing good and the other representing evil, these figures are just one example of the almost overwhelming use of symbols throughout the Aleph. The library is a different take on the opulence of the ground floor. The sleek, smooth surfaces of the lobby give way to dark hardwoods, natural fabrics and diffused light from burgundy wall lamps that give

the space a touch of "boudoir". The feeling is less nightclub and more gentleman's club, with red leather chesterfields, arrangements of blood-red roses and checkered chess-board tables, on which to place your glass of cognac. As one would expect, there's a certain Machiavellian, duplicitous air to this room, made comically clear by the fact that the only real books in it have been thrown in a bonfire heap on the floor. Holograms on the panelled walls replace real bookshelves.

The hotel's 96 rooms, deluxe rooms and suites ignore the reds and blacks of the space below them, and mix colours, textures and styles to create a bright atmosphere that, when compared

01 | The luxuriously decorated lobby with its red seating furniture is reminiscent of a boudoir.

02 | It seems as if a fire is burning in the library fanned by a pile of old books.

01   02

to the dusky richness of the hotel's public areas, can sometimes seem a little too friendly. White varnished furniture and blond wooden floor-to-ceiling sliding partition doors, that completely close off the sleeping area, reflect the daylight that pours into the room. The deep blue of the carpets that run throughout the corridors and rooms is repeated in the modern chandelier that hangs directly above the bed and in the mosaic bathrooms. With sunken bathtubs, enormous rainfall showers and the clear, sharp lines of transparent glass dividers, the Aleph's bathrooms make a crisp, refreshing environment. Bleary-eyed from a little too much Dolce Vita the night before, guests can bring themselves back

to the world of the living with revitalizing Etro toiletries, before rubbing down with a multitude of fat, plush white towels. Indeed, throughout the hotel's rooms and corridors scenes photographed from Italian life form vivid, sometimes ironic, backdrops that catch the eye: strolling at the Coliseum, people-watching in the Via Veneto, and Sophia Loren cooking spaghetti in the lift.

The hotel's paradisical spa is also a reflection of its Roman location, and the traditions of the city's most glorious and luxurious period. Modern patricians can savour the same lavish treatment as their ancestors did, in surroundings that the emperors would have felt pampered in. Sauna,

steam baths, treatments and massages are all on hand to soothe and enhance the beauty and health of the spa's fortunate guests, while a toga bar and cigar lounge cater for a very different kind of wellness.

But for the ultimate in chic profligacy, the Aleph's rooftop suites offer a private venue to beat them all. Admire the incredible vistas of Rome – complete with vintage "Martini" advertising boards – from the public terrace, then cavort, champagne in hand, in the raised jacuzzi of your own patio. Discretion goes without saying at the Aleph. The paparazzi don't stand a chance of snapping a single compromising shot.

06 07

03 | Blue carpets in the halls signify that the guests have overcome Hell.

04 | On the elevator's walls: Everyday scenes from roman life, here Sophia Loren cooking spaghetti.

05 | The furnishing style of the rooms and suites is reminiscent of the 1940s.

06 | Special comfort is offered by the bathroom. The tubs are submerged deep into the ground.

07 | Escaped from Hell again: The 99 rooms offer completely earthly comforts.

hotel de russie | rome . italy

DESIGN: Olga Polizzi, Tommaso Ziffer

A magical charm envelops the Hotel de Russie in Rome. The first image that catches the eye is a wonderful flower arrangement beneath the white vaulting against the backdrop of a terrace garden in extravagant bloom. It is hard to believe that it is just a stone's throw from the hustle and bustle of the Piazza del Popolo. Or that you just have to step through the gate - through which grand guests once passed in their carriages – to be swept up into the never-ending stream of passers-by in the Via del Babuino with its trendy designer shops. The extensive gardens, whose old pines afford ample shade, the sweet-smelling roses, orange and lemon trees, terraces on which elegant Romans meet to drink tea or their regular evening aperitif, are priceless treasures. And this in the midst of the narrow alleyways of the densely built-up centro storico in the Eternal City, where there is generally only room for a few plants in tubs.

The reopening of the Hotel de Russie in 2000 by the Rocco Forte Group was eagerly awaited in Rome as it signified the rebirth of a legendary hotel. Instead of the opulent elegance that is often found in renovated grand hotels, the designers Tommaso Ziffer and Olga Polizzi, Sir Rocco Forte's sister, decided on a unique, contemporary style that looks back on the hotel's illustrious past when famous Russians like Diaghilev and Stravinsky, or the likes of Pablo Picasso spent the night here. They mixed style elements from the dawn of the Hotel de Russie et des Iles Britanniques, as the luxury hotel was called when it opened in 1815, with those of the 1930s and 1940s. It is more reminiscent of Paris, since Rome at that time was as poor as a church mouse and costly materials were in short supply. They have cultivated a relaxed, casual chic that radiates timeless elegance but has also arrived in the here and now.

Colours are a key element in this: The lobby has been decorated in muted browns, sand and cement tones. The sofas' oversized back rests and the big leather seats invite guests to just sink into them. The modern rooms are painted a bold reed-green colour; the classical rooms are mauve combined, idiosyncratically, with old rose and marine blue. Shades of blue signal the transition to the health club, which occupies a generous 600m^2. It contains a saltwater hydropool, Turkish steam bath, jacuzzi and a fully-equipped gym which Antonio Banderas, no less, once worked out in.

01 | An oasis of calm in the middle of the centro storico in Rome: the inner courtyard of the Palazzo.

02 | Small, but exquisite: the hydro pool in the health club decorated with blue mosaics.

03

04

03 | Contemporary art by the entrance to reception creates a link to modern-day Rome.

04 | A winking churchman watches over the goings on in the lobby.

05 | In a niche in the wall: "Man and Woman" (1999) by the Italian artist Oliviero Rainaldi.

06 | The lobby has a relaxed, contemporary feel.

07 | Strategically placed objects are eye-catching.

12 13

08 | Most of the 129 rooms in the hotel look out onto the inner courtyard and the Mediterranean gardens.

09 | A mix of styles: the Directoire chair is from France, the mirror is based on 1920s design and the photograph next to the bed is taken from Robert Mapplethorpe's flower series.

10 | Carefully sculpted box trees surrounding the parapet.

11 | The calming sound of water splashing in the fountain to accompany evening meals

12 | The rates are exclusive, but guests find spacious and luxury rooms.

13 | One of the suites in the sixth floor, which also gives access to the roof-top terrace with a spectacular view onto the city.

grand resort lagonissi | athens . greece

DESIGN: Pantellis Mantonanakis

This large resort along the sea front is only a 15-minute drive from Athens airport and provides a wide variety of rooms, suites, restaurants, bars and clubs. Guests have a choice of four standards of accommodation ranging from Comfort Club, to Premium and Exclusive, to the so-called Platinum Club for those who are not troubled by such details as credit limits. What all the rooms have in common is colourful, though not garish fittings and furnishings and, depending on the category, ample to extravagant amounts of space with many added extras. The Vista Suite, for example, in the most basic category, stretches over a total of 42 m², comprising of a bedroom, a living room, a large, marble-floored bathroom and a private balcony. Luxury is piled upon luxury as one moves up the categories to the superlative 380 m² Royal Suite boasting two large bedrooms, a living room with an open fireplace, a large-screen TV and stereo system with Dolby surround,

two bathrooms with separate shower (including a mini-TV), quarters for a personal butler, a gymnasium with steam bath and massage table, a fully-furnished kitchen, separate work and office space, an indoor and outdoor pool and private parking – not to mention the breathtaking view across the sea from the balcony at sunset.

Guests of a sporty disposition have a choice of how and where to let off steam: on one of the flood-lit tennis courts, playing volleyball or in the heated swimming pool. Sun-worshippers can recline on the soft sand along one of the 16 beaches belonging to the hotel. Hungry? There are various restaurants and bistros in the grounds serving everything from traditional Greek cuisine in the "Ouzeri", to Italian cooking in the Captain's House, to Polynesian delicacies in the Kohylia Restaurant. The latter also serves sushi specialities.

The perfect, unobtrusive all-round service leaves nothing to be desired: staff will gladly find you a sunshade, dry-clean your soiled clothes or put a new tint of colour in your hair. Guests will find someone to deal with any problem in the blink of an eye, so that nothing can come between you and total relaxation. Need a babysitter? Nothing could be simpler. The babysitting service, along with the many other special facilities provided for children (play area, childminders, minigolf), take the strain off family holidays too. The complex is not only perfect for a relaxing break, but also has conference facilities. Rooms facing the sea with natural lighting are available for meetings of five to 300 people. If required, exhibition space of up to 250 m² can be hired and set up for company or product presentations.

01 | With a total of six restaurants there is hardly a chance of getting bored even after a week.

One of the design highlights is the Polynesian restaurant with Sushi Bar "Kohylia".

02 | View onto the private pool in the living room in the Royal-Suite.

03 | Bathroom in a Junior-Suite.

04 | Bedroom with dome of light.

05 | Dining room in Royal-Suite.

06 | The building consists of 188 rooms, suites and bungalows,
22 of which have a private pool.

04 05

06

belvedere | mykonos . greece

DESIGN: Paris Liakos, Agelos Agelopoulos

This hotel combines the picturesque charm of the old Aegean with the jet-set, artsy lifestyle of the Mykonos School of Fine Arts District. Owners Tasos and Nikolas Ioanidis have updated an old mansion located just minutes away from boutiques and bars, which have made the area popular all year round.

The Ioanidis family has owned this luxurious address for years. It is independently owned and operated, and not affiliated to any chain. This guarantees that both the hotel's ambience and highly personal level of service are unique. Many guests are regulars of the hotel's 41 rooms and six suites, which are scattered between six separate poolside complexes. There are also six suites, and a large VIP apartment.

Architecture includes all of the features, which make the Cycladic style unique: whitewashed

interlocking cubes, curved walls, airy veranda and simple wooden roofs. Rooms are individually appointed, but all feature sailcloth blinds to shield the blazing sun, lots of pillows for lounging and white marble bathrooms.

Recreation at the Hotel Belvedere means relaxing with a book at the poolside, or simply gazing at the magnificent Aegean sprawling on the horizon. For the somewhat more ambitious, there is a fitness studio in the hotel's spa area, which also features a jacuzzi and steam bath. But most people come to the Belvedere for socializing, and there are plenty of opportunities around the pool. There are two separate bars, a convivial fireplace in the lounge, and a restaurant with a huge wine list.

01 | This private open Lounge with a view over the Mykonos town counts as one of the most desired living areas on the island .

05 06

07

02 | Brilliant white dominates the architecture even in the inner rooms.

03 | Stairs, paths and small lanes connect the six buildings in the grounds.

04 | Numerous decorative details can be discovered when walking around the grounds.

05 | Even the roof landscape can be partly accessed.

06 | In total there are 41 rooms and six suites, several of which with such terraces.

07 | The origins of some hotel buildings reaches as far back as 1850.

08  09

10

08 | White marble decorates the bathroom.

09 | For those who want to – can just pull the sail across.

10 | One can retire on such day beds at the pool.

11 | Less furniture – more space for relaxation.

12 | The family Ioanidis, who own the hotel area, have successively
     had it renovated and extended since its opening in 1995. One of
     the latest additions is the Sunset Bar.

katikies | santorini . greece

DESIGN: Ilias Apostolidis, Giorgos Lizardos, Nikos Tzelepis

Santorini is perhaps the best known of Greece's Cycladic Islands, and steeped in myths. It was here that an ancient civilization sunk into the sea, possibly giving rise to the story of the lost city of Atlantis. Today the name Santorini stands for romantic associations all its own; white cottages with blue roofs, stairways cut into rock, and tiny chapels surrounded everywhere by shimmering seas. The island itself is almost too beautiful to be believed. And on its most western point, in the village of Oia, is a hideaway, which unites the breathtaking vistas of Santorini with the amenities of a luxury hotel.

The Hotel Katikies provides experiences that live up to the reputation of peace and sensuality associated with the story of Atlantis. As twilight envelopes the surrounding seas, the air cools, and pools of warm effervescent water become illuminated in bright green. Your own private

masseuse, clad in white, approaches to knead and invigorate tired muscles with precious oils. Nothing unusual about this here, poolside massages are only part of the service at Hotel Katikies.

One of its most distinguishing features is the terraced design, with rooms, pools and niches on many levels. It's like a tiny whitewashed village set on a cliff, with winding stairs and passageways everywhere. Dazzling blue waters lap the rocks 100 meters below. Lazy yachts cruise the sea, and in the early evening, the sunset is dramatic.

All rooms are traditionally furnished with antique chairs, mirrors, sidetables and lamps. Some suites feature vaulted ceilings, giving the space the feel of a cozy grotto. Bathrooms, sitting rooms, and verandas are outfitted with modern

conveniences. Selected suites also have private jacuzzis. The hotel has three separate dining opportunities: one at the pool, another in an intimate grotto, and finally an open-air gourmet restaurant. A selection of fine wines, from vineyards on Santorini's rich volcanic soil, are available for tasting at sunset.

01 | Just like a small fishing village: the hotel area snuggles into the hang of Oia at the northern tip of the island.

02 03
04

02 | Typical Greek postcard idyll. The 34 rooms and 15 suites are connected over numerous steps on the inside and outside.

03 | The edge of the pool is completely flat here but so low that the optical transition to the sea is mainly determined by various blue tones.

04 | View over a larger roof terrace on the Aegean Sea...

05 | ...and from a smaller one.

06 | Almost all the rooms have such terraces.

07 | One mainly lives in the open air. One mainly sleeps in comfortable rooms reminiscent of caves.

05

06 07

anassa | polis . cyprus

DESIGN: James Northcutt Associates, L.A.

Island of the Gods... that's what tourist brochures call this sun-drenched island in the southeastern Mediterranean. The best known legends about Aphrodite, the goddess of love, are associated with Cyprus. Hesiod says she was born out of the foam at the rocky cliffs near Paphos. And in a secluded bay in the northwest, Aphrodite met Adonis, her lover. Not far from these myth-inspiriting sites lies the modern luxury resort of Anassa. The idyllic setting is an archetype of a Mediterranean dream. Anassa is located on a bay whose beaches have sand as fine as in any hourglass. It's surrounded by soft cliffs, and close to the wild landscape of the Akamas Peninsula. And the slightly elevated location of the resort provides breathtaking views of the deep-blue waters of the Chrysochou Bay.

The resort complex blends perfectly with the landscape, resembling a Cypriot village. The brilliant whitewashed structures are topped by clay tiles and surrounded by opulent gardens of fragrant bourgainvillea, jasmine and lavender. Shutters on the windows and doors are a calming turquoise colour. There is a square surrounded by arts and crafts workshops, and the village chapel can used by modern lovers for weddings. Refreshment in the local café is offered in the form of ouzo, the traditional anise-flavoured spirit. Cyprus has seen many conquerers in its 8,000-year history. The Egyptians, the Greeks, the Romans... and each have left traces of their own unique cultures. These are reflected in the interior design by James Northcutt Associates. The magnificent cupola in the expansive lobby area, the Roman mosaics, frescos inspired by Venetian artists. The garden decorated with ancient amphora, clay containers for storing oil, olives and wine that were ubiquitous in antique times. The 184 rooms at Anassa feature muted, natural tones. Their cream-coloured marble floors and delicate curtains combine to complete the unencumbered and airy atmosphere.

The wellness area is inspired by Roman baths. Fragrant breezes come into the bright swimming pool area when the windows are left open. There is a sea water tub with powerful jet nozzles for underwater massage. Anassa has two separate saunas, and two steam baths as well. The spa specializes in Thalasso therapy, which harnesses the regenerative secrets of the sea: algae, mud, salt and sand. Thalasso comes from the Greek word for "ocean," and its healing powers have been exploited since ancient times. At Anassa, Thalasso therapy is popular for relieving stress, stimulating circulation, and restoring the body's natural balance... one of the prime principles of ancient Greek thinking.

02 | 03

02 | The whirlpool has shade from a pergola in the Mediterranean style.

03 | The eleven garden suites each have a private pool on the terrace.

04 | The sunlight-flooded swimming baths in the wellness area measures eight times eighteen metres.

05 | Turquoise blue lamella window shutters and white-dipped walls characterise the Mediterranean flair.

04 05

06 | The stairs curves imposingly down into the Lobby.

07 | Quiet and comfort is offered by the generously sized suites which
are furnished in soft beige and crème tones.

08 | Four-toaster with a view – the Mediterrean Sea is just behind the
terrace.

09 | Dining room with rough stone flooring.

10 | Light and shadows on classicals archs.

08

09 10

marmara | bodrum . turkey

DESIGN: Christian Allart, Ersen Gürsel

The intense deep blue of the bay is shot through with streaks of soft violet. Yachts – some extremely grand – gather at the harbour as night descends over the town. A parade of bright white houses shimmers in the light of the setting sun; lanterns, smouldering like hot ashes, weave abstract patterns. Centre stage, a mosque's twin minarets. A picture of peace.

This is the view from the terrace of the Marmara Hotel in Bodrum, a first-class destination in the deep south of the Turkish riviera. Each evening, this ancient port at the southern end of the Aegean Sea springs into gear. Renowned for its nightlife, the town is packed with seaside bars and restaurants where the party lasts well into the small hours. The hotel, though, appears to have carved out a more exclusive location for itself: like an observation platform, perched high on a hill, well away from the hurly-burly and surrounded by gardens and nature, with breathtaking views across the bay.

During the day, an aesthetic full of light unfolds throughout the hotel. All the rooms, mainly in white, are "composed" around the magnificent panorama. Whether you open your shutters, step onto the terrace from one of the six garden suites or the exclusive Party Animal Suite, laze around in the pools on the sun patio or just gaze out from the spa with its jacuzzi, hammam and gym – everywhere you look there is the same distant vista. The rough sandstone walls, reminiscent of the ancient ruins in the area, add their own touch of the past.

French interior designer Christian Allart has given the hotel an elegance that is simplicity itself, whilst the interior – with its many Middle Eastern treasures – exudes a Mediterranean air.

The Turkish bath is entirely marble; the lobby has enormous amphoras, silk armchairs and frog-style coffee tables. The design of the rooms has an almost Oriental rigour. Taste pared down to the essentials. Around a fireplace in the lounge, a collection of old tools is a reminder of the lifestyle of local people in days gone by.

The region's fish and plentiful fruit make for a cuisine that is light and healthy. At the hotel restaurant, a French and a local chef serve fine fusion food throughout the day. The perfect way to round off days of revitalisation, endless bathing and diving along the coast – or maybe an excursion or tour through the mountains inland – is a "candlelight finale" with a fruity cocktail on the terrace. Still, nobody would be in the least surprised if guests never felt the need to venture beyond this luxury hotel at all.

01   02

01 | The first -class wellness destination is situated at the
     southern most point of the Turkish Riviera.

02 | Lobby with reception in the background, in-between this
     golden amphora and silk armchairs.

03 | Architectonic mixture of plain high buildings and eclectic interior.

04 | Playful details such as the frog-like couch table is design to loosen the senses.

05 | Pool and terrace with view onto Bodrum, known for its fantastic nightlife.

06 | The complex never looks too full with its 94 rooms and 6 suites.

07 | Details, accessories and art objects are to be highlighted. Architectonic shapes, wall and floor colouring are therefore in the background.

05
06 07

hillside su | antalya . turkey

DESIGN: Eren Talu, Yael Bahor, Asli Eke, Merve Yoneyman

A fusion of cutting edge design? Shocking like Schrager? Or a disco destination with five hotel stars? Opinions of the Hillside Su Hotel could hardly be more conflicting. The way this new "city resort" is striking a pose bears no relation to the softly-softly style of architecture that attempts to blend in with local contexts and traditions. A seven-storey, sheer concrete block, eminently sleek but very muscular, soars upwards with swanky coloured lights announcing, "Here I am".

"How did we end up in Las Vegas?" might be an initial reaction. No matter: the Turkish riviera and all its attractions – long beaches, orange groves, jagged cliffs and swathes of coastline – is on the doorstep. Is that not a contradiction? Absolutely. But the contrast is no mere accident of design. For ten years, the creative people and managers at Hillside Leisure Group pondered over this project, rejecting designs one after the other

until they hit on what appeared to be the perfect concept. Eren Talu's work is a challenge to conventional expectations. Take the lobby, for instance: the massive mirror balls suspended more than twenty metres from the ceiling bathe the balconies of all seven floors in a hail of shimmering stars, enhanced by the red and blue lighting effects framing the scene.

This interplay between the white architecture, the mirrors and the lighting effects is a feature throughout the 294 apartments and 41 suites. People who like their hotels to be cosy would probably be happier elsewhere. Even the guest rooms and the superior suites here convey the feeling of being on a cruise ship gliding through outer space. Everything is white, even the TV. The acres of soft upholstery and intimate seating areas in white leather – what else? – are particularly inviting places to chill out. Coolness

is very definitely the key, with vast mirrors on the walls not only opening up the room but also allowing guests to analyse their every move and every aspect of their appearance – and make readjustments as necessary. This colourless space landscape is then shrouded in sensuous stratospheres by lava lamps and light installations, casting it adrift from planet Earth.

The hotel exudes an atmosphere full of buzz and energy, as well as offering a whole host of recreational activities for any guests needing to recharge their batteries after indulging a little too eagerly. There are indoor and outdoor heated swimming pools, tennis and squash facilities, and a spa providing a range of therapies for hard-pressed, stressed-out bodies. But Eren Talu has one more ace up his sleeve: as you sink down into the shallow pool water, you are enveloped by gentle sounds beneath the surface.

01 | Room with a view: A cool room ambience melts into the colours of the breaking night.

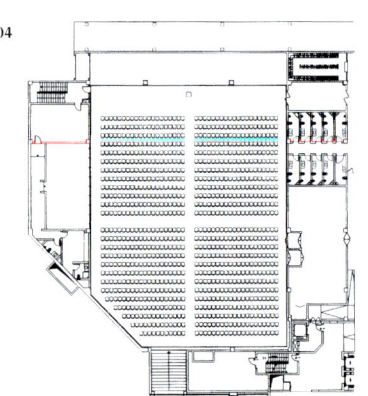

04

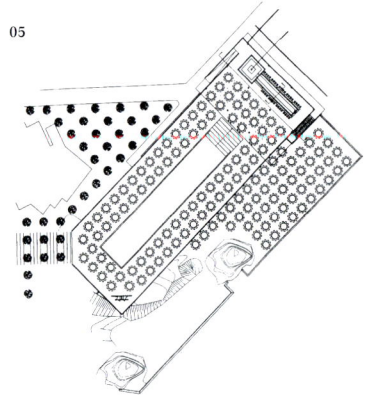

05

02 | Everything in the room is white if there were no light installations. In this way dreams gain colour.

03 | Leather upholstery, a bed like a field. Mirror over mirror. Room like a spatial landscape in which the future has already begun.

04 | 05 The plans give an idea of the size of the complex.

02   03

06 07

08

06 | Everything is white in the bathroom as well. What a surprise!
Pricking essences provide refreshing.

07 | Rooms that transport breadth. The mirrors inside open up
several perspectives.

08 | The interplay between effects of light and modern white
architecture, exhibited in the Lobby par excellence.

hotel summary

Country / Location		Address	Information	Architecture & Design	Page
Portugal	Madeira	Choupana Hills Resort & Spa Travessa do Largo da Choupana 9050-286 Funchal Portugal www.choupanahills.com	opened 2002 60 Deluxe rooms, 4 suites restaurant, bars, lounge, fitness- and spa center, turkish steam bath, sauna, jacuzzi, indoor and outdoor pool, nearby 2 golf courts (18 and 27 holes). On the hills over Funchal.	Michel de Camaret Didier Lefort	8
Portugal	Madeira	Crowne Plaza Resort Estrada Monumental, 175-177 9000-100 Funchal Portugal www.madeira.crowneplaza.com	opened 2000 276 rooms, 24 suites restaurants, conference rooms for up to 200 people, Thalasso therapy in the Thalgo Marine Spa, PADI diving center. 20 min. walk from Funchal city's center.	Ricardo Nogueira Caldeira Silva	12
Portugal	Madeira	Estalagem da Ponta do Sol Quinta da Rochinha 9360-121 Ponta do Sol Portugal www.pontadosol.com	opened 2001 54 rooms snack bar, restaurant, lounge bar, swimming pool, heated indoor pool, sauna, jacuzzi, fitness, sundeck with panoramic view. At Madeira's southern coast, 30 min. drive to Funchal.	Tiago Oliveira	16
Portugal	Madeira	Quinta da Casa Branca Rua da Casa Branca, No. 7 0000 000 Funchal Portugal www.quintacasabranca.pt	opened 1999/2001 29 rooms, 12 Deluxe rooms and 2 suites 2 restaurants, 2 bars, „Al Fresco" spa with jacuzzi, sauna, turkish steam bath. In the center of Funchal, surrounded by wonderful gardens.	Joao Favila V.S. Menezes Teresa Gois Ferraira Luis Rosario	22
Spain	Gran Canaria	Palm Beach Avenida de Moya 8 35100 Playa del Inglés Gran Canaria Spain www.hotel-palm-beach.com	opened 2002 327 rooms and suites restaurants „La Trattoria", „Orangerie", „Seaside Barbecue", „Snack Bar", „Africana Pool Bar", „Salon Bar", variety of sports and wellness offerings, hammam. 38 km from the airport.	Alberto Pinto	26

Country / Location	Address	Information	Architecture & Design	Page
Spain　　　La Parra	Hospederia Convento de La Parra Santa Maria, 16 – La Parra 06176 Badajoz Spain www.laparra.net	opened 2002 15 double rooms, 4 single rooms, 2 junior suites In a former monastry, complete absence of intrusive elements such as TV or radio.	Francisco Wenao Maria Ulecia Xavier Munier	30
Spain　　　Mallorca	Hotel Maricel Carretera de Andratx, No. 11. Cas Catalá 07184 Calvià, Palma de Mallorca Spain www.hospes.es	opened 2003 24 double rooms and 4 suites restaurant, terrace with sea view, library, pool and wellness area, private landing place. 15 min. drive from airport.	Xavier Claramont	34
France　　Bordeaux-Martillac	Les Sources de Caudalie Chemin de Smith Haut-Lafitte 33650 Bordeaux-Martillac France www.sources-caudalie.com	opened 1995 49 rooms and suites 2 restaurants, cigar club, bar, spa area, vinotherapy, swimming pool, fitness.	Yves Collet	38
France　　　Hagetmau	Hôtel des Lacs d'Halco Route de Cazalies 40700 Hagetmau France www.hotel-des-lacs-dhalco.com	opened 2001 24 rooms restaurant „La Dame du Lac", regular tasting sessions, indoor pool, swimming in the lake. At the Pyrenees, 1 hr. drive to the atlantic ocean.	Eric Raffy	42
France　　　Marseille	Sofitel Palm Beach 200 Corniche J.F. Kennedy 13007 Marseille France www.accorhotels.com	opened 2002 160 rooms, including 15 suites restaurant with Mediterrean cuisine, 6 meeting rooms, ball room for up to 340 people, fitness center, sauna, jacuzzi, swimming pool, private beach.	Claire Fatosme Christian Lefèvre	46

hotel summary

Country / Location		Address	Information	Architecture & Design	Page
United Kingdom	Cardiff	The St. David's Hotel & Spa Havannah Street Cardiff, CF10 5SD United Kingdom www.thestdavidshotel.com	opened 1999 132 rooms, including 19 suites 7 meeting rooms, restaurant, spa dining suite, „Tides Bar", variety of spa offerings including hydrotherapy, aromatherapy.	Olga Polizzi	50
United Kingdom	Somerset	Babington House Babington BA11 3RW United Kingdom www.babingtonhouse.co.uk	opened 1999 27 rooms restaurants, terrace, bar, meeting rooms, screening room, spa area with treatments in cowshed or tee pee, fitness, sauna, 2 pools, tennis court, crocket field. Nearby Bath, 180 km from London Heathrow.	Simon Morray-Jones	54
United Kingdom Gloucestershire		Cowley Manor Cowley GL 53 9NL United Kingdom www.cowleymanor.com	opened 2002 30 rooms restaurant, spa area with sauna, steam bath, indoor and outdoor swimming pool.	Ryan de Matos Storey	58
United Kingdom	London	One Aldwych 1 Aldwych London WC2B 4BZ United Kingdom www.onealdwych.com	built 1907, reopended 1998 105 rooms and suites restaurants, bars, pivate screening room for up to 30 people, 2 meeting rooms, state-of-the-art fitness center, swimming pool with underwater music. In the heart of London.	Mary Fox-Linton Gordon Campbell Gray	62
Großbritannien	London	Sanderson Agua Spa Berners Street London W1P 3 AD United Kingdom www.ianschragerhotels.com	opened 2000 150 rooms, 2 penthouse apartments restaurant directed by Alain Ducasse, 10.000 m^2 spa area on two levels with health, fitness, beauty and wellness center. In heart of Soho, London's media and cultural center.	Philippe Starck Michael Nash Associates	66

Country / Location		Address	Information	Architecture & Design	Page
United Kingdom	Yorkshire	Seaham Hall	opened 2001	Nappers Architects	70
		Lord Byron's Walk	18 suites, 1 penthouse		
		Seaham SR7 1AG	4 meeting rooms, ballroom for up to 100 people,		
		United Kingdom	oriental spa area, pool, sauna, jacuzzi,		
			massage and beauty facilities.		
		www.seaham-hall.com	At England's east coast between York and Edinburgh.		
Germany	Helgoland	Atoll	opened 1999	Alison Brooks	74
		Lung Wai 27	43 double rooms, 7 suites, 1 apartment		
		27498 Helgoland	restaurant, bistro, café and bar, 2 meeting rooms,		
		Germany	wellness area with swimming pool, sauna, fitness.		
			At the southern pier of Helgoland.		
		www.atoll.de			
Germany	Berlin	Grand Hyatt	opened 1998	José Rafael Moneo	78
		Marlene-Dietrich-Platz 2	326 rooms, 16 suites, including 2 president's suites	Hannes Wettstein	
		10785 Berlin	3 restaurants,		
		Germany	7 meeting rooms, ball room for up to 850 people,		
			Club Olympus Spa & Fitness.		
		www.berlin.hyatt.de	In the heart of Berlin at the Potsdamer Platz.		
Switzerland	Gstaad	Grand Hotel Bellevue	reopened 2002	Gottfried Hauswirth	82
		Hauptstrasse	35 rooms, including 3 suites	Theo Jakob AG	
		3780 Gstaad	restaurants „Coelho" and „Prado", bar „Bellevue"		
		Switzerland	wellness for body and mind.		
			In the center of a 18.000 m^2 park.		
		www.bellevue-gstaad.ch			
Switzerland	Lenk	Lenkerhof Alpine Resort	reopened 2002	Rolf Balmer	86
		Postfach 241	33 rooms, 31 junior suites, 4 suites		
		3775 Lenk im Simmental	meeting room for up to 120 people,		
		Switzerland	restaurants, bar, vinotheque,		
			wellness at 2000 m above sealevel.		
		www.lenkerhof.ch	In a small valley, approx. 20 km from Gstaad.		

hotel summary

Country / Location	Address	Information	Architecture & Design	Page
Switzerland — Weggis	Parkhotel Weggis Hertensteinstrasse 34 6353 Weggis Switzerland www.phw.ch	new building opened 2002 34 rooms, 9 suites restaurants, bars, vinotheque, wellness area with six 60 m^2 „spa cottages", courtyard with japanese garden.	Vincenz Erni Vadian Metting van Rijn Müller und Partner AG Jörg Rickli	90
Switzerland — Vals	Therme Vals 7132 Vals Switzerland www.therme-vals.ch	opened 1998 46 four-star-rooms, 80 three-star-rooms fresh, creative cuisinge in different restaurants, thermal bath, variety of wellness offerings such as ayurvedic synchron massage, shiatsu massage, stone massage. At the Val Lumnezia, 45 min. drive from Chur.	Peter Zumthor	96
Austria — Dornbirn	Rickatschwende Boedelestrasse 6850 Dornbirn Austria www.rickatschwende.com	opened 2000 49 rooms Gault-Millau-awarded gourmet restaurant, wellness area, sauna, steam bath, Kneipp therapies.	Thomas Hämmerle, Huber Planungsbaugesellschaft Marika Marte, Enea	100
Austria — Bezau	Post Hotel Brugg 35 6870 Bezau Austria www.hotelpostbezau.com	new building opened 1998 53 rooms 3 meeting rooms, Gault Millau awarded restaurant, beauty and wellness area, traditional chinese medicine.	Oskar Leo Kaufmann Johannes Kaufmann	104
Austria — Ischgl	Design Hotel Madlein 6561 Ischgl, Tirol Austria www.ischglmadlein.com www.designhotels.com	new building opened 2000 80 rooms restaurant and „Pacha-Ischgl", „Coyote Ugly Bar", „Sushi Lounge", open fireplace, zen garden, spa area with Thalgo, Thalasso, anti-aging programmes. Just a step away from the best ski areas.	Sabine Mescherowsky Gregor Mescherowsky	108

Country / Location		Address	Information	Architecture & Design	Page
Austria	Kitzbühel	Schwarzer Adler	reopened 2001	Wolfgang Pöschl	114
		Florianigasse 15	83 rooms	Kay Sperling	
		6370 Kitzbühel	Gault-Millau-awarded restaurant,	Daniela Flamm	
		Austria	futuristic wellness area „Black Spa".		
		www.adlerkitz.at			
Italy	Lana	Vigilius Mountain Resort	opened 2003	Matteo Thun	118
		Monte San Vigilio 1500	31 rooms, 8 suites	Christina von Berg	
		39011 Lana / South Tyrol	flexible meeting rooms for 40-80 people,		
		Italy	restaurant, „Stube", wine cellar,		
			spa, indoor pool, whirlpool, sauna, steam bath.		
		www.vigilius.it	1800 m above sea level, 15 km from Merano.		
Italy	Rom	Aleph	opened 2003	Adam D. Tihany	122
		Via di San Basilio 15	99 rooms, including 2 suites		
		00187 Rome	restaurant, bar & wine lounge, library lounge,		
		Italy	spa area with turkish steam bath, sauna, pools.		
			In the heart of Rome, 5 min. walk from „Termini" station.		
		www.aleph.it			
Italy	Rom	Hotel de Russie	opened 2000	Tommaso Ziffer	126
		Via del Babuino 9	129 rooms, including 25 suites	Olga Polizzi	
		00187 Rome	5 meeting rooms, business center,		
		Italy	restaurant „Le Jardin de Russie", „Stravinskij Bar",		
		www.hotelderussie.it	garden terraces, health club and spa.		
		www.roccofortehotels.com	At the Piazza del Popolo, nearby the Spanish Steps.		
Greece	Lagonissi	Grand Resort Lagonissi	opened 2003		132
		40km Athens-Sounio Ave.	138 rooms, 50 suites		
		19010 Lagonissi	7 different restaurants, bars, clubs,		
		Greece	meeting rooms for up to 300 people.		
			40 km from Athens, located directly by the sea.		
		www.lagonissiresort.gr			

hotel summary

Country / Location		Address	Information	Architecture & Design	Page
Greece	Mykonos	Belvedere	opened 1995	Paris Liakos	136
		School of Fine Arts District	41 rooms, 6 suites, VIP suite	Agelos Agelopoulos	
		84600 Mykonos	restaurants, snack and pool bar, sunset bar,		
		Greece	fitness studio, swimming pool, steam bath, internet room,		
			CD and DVD library.		
		www.belvederehotel.com	5 min. walk from the glamorous island's center.		
Greece	Santorin	Katikies	reopened 2000	Ilias Apostolidis	142
		847 02 Oia	34 rooms and 15 suites	Nikos Tzelepis	
		Santorini	restaurants: „Karini" (pool), „Katikies Gourmet", „White Cave",	Giorgos Lizardos	
		Greece	2 swimming pools, wellness facilities, water sports.		
		www.katikies.com			
Cyprus	Polis	Anassa	opened 1998	James Northcutt Associates	146
		P.O. Box 66006	184 rooms and suites	Henri Del Olmo	
		8830 Polis	4 restaurants, bars, terraces,		
		Cyprus	roman bath, seawater pool, Thalasso treatments.		
			At the outskirts of Akamas peninsula		
		www.thanos-hotels.com.cy			
Turkey	Bodrum	Marmara	opened 2000	Ersen Gürsel	152
		Pk 199	100 rooms, including 6 suites	Christian Allart	
		48400 Bodrum	meeting rooms, „Tuti" restaurant, lounge bar, pool bar,		
		Turkey	fitness center, 2 swimming pools, squash courts.		
			With a magnificent view on Bodrum and its medieval castle.		
		www.themarmarabodrum.com			
Turkey	Antalya	Hillside Su	opened 2003	Eren Talu	156
		Konyaalti	294 rooms, including 41 suites	Yael Bahior	
		07050 Antalya	ball room, 8 meeting rooms,	Asli Eke	
		Turkey	buffet and á la carte restaurants, lounge, bars,	Merve Yoneyman	
			heated indoor and outdoor pools, spa area.		
		www.hillside.com.tr	1,5 km from the center of Antalya.		

architects & designers

photo credits

All other photos by: Martin Nicholas Kunz

imprint

Bibliographic information published by Die Deutsche Bibliothek
Die Deutsche Bibliothek lists this publication in the Deutsche
Nationalbibliografie; detailed bibliographic data are available in
the Internet at http://dnb.ddb.de

ISBN 3-929638-85-1

Copyright © 2003 Martin Nicholas Kunz
Copyright © 2003 lebensart global networks AG, Augsburg
Copyright © 2003 avedition GmbH, Ludwigsburg
All rights reserved.

Printed in Germany

Publisher | Martin Nicholas Kunz
Translations: Lingserve, Rachel Arnold, Kevin Cote,
Vineeta Manglani, Scott M. Crouch
Copy editing: Vineeta Manglani
Texts (page) | Ursula Dietmair (26, 118); Sybille Eck (30, 42, 96,
138); Bärbel Holzberg (38, 82, 86, 90, 100, 104, 108, 114, 122,
126, 146); Ina Sinterhauf (78); Heinfried Tacke (34, 46, 50, 58,
142, 148, 152, 156); Martin Kunz (8, 12, 16, 22, 54, 66, 70, 74)
Research | Scott Michael Crouch, Bärbel Holzberg,
Katja de Marné, Ulrike Paul
Art Direction | Willem Krauss, Michael Schickinger
Production | Markus Hartmann
Printing | Leibfarth & Schwarz GmbH & Co KG

lebensart global networks AG
Konrad-Adenauer-Allee 35 | 86150 Augsburg | Germany
http://www.lebensart-ag.com | publishing@lebensart-ag.com

avedition GmbH
Königsallee 57 | 71638 Ludwigsburg | Germany
p +49-7141-1477391 | f +49-7141-1477399
http://www.avedition.de | info@avedition.de

Special thanks to: C. Abbadie, Les Sources de Caudalie |
Sabine Alge, Rickatschwende | Günther Aloys, Design Hotel
Madlein | Beatrice & Georges Ambühl, Grand Hotel Bellevue |
Barbara Anklin, Hotel de Russie | Maria Jesús Asiain, Maricel |
Domenico Basciano, Sofitel Palm Beach | Louise Belcher, The
St. David's Hotel & Spa | Romy Bitschnau, Post Hotel |
Alexandra Boge, Rickatschwende | Svenja Büschung, Palm
Beach | Gordon Campbell-Gray, Parkhotel
Weggis | Elizabeth Crompton-Batt, Cowley Manor | Dinitra
Daskalaki, Grand Resort Lagonissi | Michael Di Lonardo, Vigilius
Mountain Resort | André Diogo, Estalagem da Ponta do Sol |
Asli Eke, Hillside Su | Pauline Engelse, Crowne Plaza Resort |
Isabel F. Ferrez, Quinta da Casa Branca | Rocco Forte, Rocco
Forte Hotels | Bruno Franchi, Studio Matteo Thun | Philippe
Frutiger, Lenkerhof Alpine Resort | Dimitris Georgoulis,
Belvedere | Gregor Gerlach, Seaside Hotels | Katja Hasselkus,
WeberBenAmmar PR | Katja Hekkala, Choupana Hills | Alice
Hemmer, Schubert & Schubert, Hamburg | Michael Hönigmann,
Schwarzer Adler | Nicole Hottinger, Lenkerhof Alpine Resort |
Freddy Hürst, Grand Hyatt Berlin | Tasos and Nikolas Ionnidis,
Belvedere | Aylin Kaltakkiran, Marmara | Claudia Knapp, Therme
Vals | Reto Kocher, Grand Hotel Bellevue | Corinne Krause,
Bela Kosmetik | Gill McFadden, Seaham Hall | Galia & Thanos
Michaelides, Anassa | Estelle Monti, Sofitel Palm Beach |
Philippe Moureau, Choupana Hills | Ignacio Pérez, Maricel |
Frantzeska Polikandrioti, Belvedere | Wolfgang Pöschl,
Schwarzer Adler | Eric Raffy | Kerstin Riedel, Grand Hyatt Berlin |
Howard Rombough, One Aldwych | Waltraud Rüf, Hotel Post |
Katrin Schulz, Schubert & Schubert, Hamburg | Kay Sperling,
Schwarzer Adler | Gilles Stellardo, Aleph | Thomas Straumann,
Grand Hotel Bellevue | Eren Talu, Hillside Su | Astrid Thamm,
Atoll | Matteo Thun | Murat Tufan, Hillside Su | Eleni Tzanou,
Katikies | Neslihan Ugurlu, Hillside Su | Marleen Vietzke, Studio
Matteo Thun | Christina von Berg, Studio Matteo Thun | Lisa
Walker, Purple PR | Barbara Widera, Babington House | Jasmina
Ziouziou, Palm Beach | Annalisa Zumthor, Therme Vals

design hotelswww.designhotels.com
Small Luxury Hotelswww.slh.com

Martin Nicholas Kunz

Born 1957 in Hollywood.
Martin is Senior Vice
President publishing and
communications of lebensart
global networks AG. Martin
worked as an editor for
several German and other
international magazines such
as "design report" and was
Managing Director of New
Media for the German
publisher DVA, a company
known for its architecture,
design and craft books,
magazines and web sites.
He is author and co-author
of several design, craft and
construction books, and
since 2001, author and
publisher of the avedition
lebensart book series "best
designed…"; seven books of
the world's most beautiful
hotels.